Offshore Wind Turbine Technology

MingWei Ge · Li Li · Hang Meng · Kai Long

Offshore Wind Turbine Technology

中国工信出版集团 電子工業出版社 PUBLISHING HOUSE OF ELECTRONICS INDUSTRY http://www.phei.com.cn

Springer

MingWei Ge
New Energy School
North China Electric Power University
Beijing, China

Li Li
New Energy School
North China Electric Power University
Beijing, China

Hang Meng
New Energy School
North China Electric Power University
Beijing, China

Kai Long
New Energy School
North China Electric Power University
Beijing, China

ISBN 978-981-96-2174-3 ISBN 978-981-96-2175-0 (eBook)
https://doi.org/10.1007/978-981-96-2175-0

Jointly published with Publishing House of Electronics Industry.
The print edition is not for sale in China (Mainland). Customers from China (Mainland) please order the print book from: Publishing House of Electronics Industry.

This Springer imprint is published by the registered company Springer Nature Singapore Pte Ltd.
The registered company address is: 152 Beach Road, #21-01/04 Gateway East, Singapore 189721, Singapore

Preface

The vigorous development of wind power generation (referred to as wind energy) is a key strategy in China's energy transition and an essential pathway for the country to achieve its "30·60" carbon peaking and carbon neutrality goals. Carbon peaking and carbon neutrality represent a broad and profound transformation of the economic and social systems. It is foreseeable that in the coming decades, the technological structure, industrial structure, and development model of China's industry will undergo significant changes. Wind power is a highly important form of utilization among renewable energy sources. More than 90 countries worldwide have already undertaken large-scale development and construction of wind power generation. Since 2010, China's installed wind power capacity has surpassed that of the USA, making it the country with the largest installed wind power capacity in the world. By the end of 2020, the cumulative global installed wind power capacity exceeded 733 GW, with China accounting for more than 35% of the total. In 2020, the global newly installed wind power capacity reached 111 GW, with China contributing nearly 65% of the total.

In recent years, China's offshore wind power has developed rapidly, with numerous offshore wind farms commencing construction and several large-scale wind turbines being put into operation. Compared to onshore wind power, offshore wind power operates in more complex environments, faces greater construction challenges, and incurs higher operation and maintenance costs. In addition, limiting factors such as the relatively poor quality of offshore wind resources in China (low average wind speeds and high extreme wind speeds), and the low reliability of wind turbines and key components, result in a higher levelized cost of electricity. Against this backdrop, this book focuses on offshore wind turbines and the core power generation equipment and addresses the unique wind-wave-current coupling environment and offshore geological conditions. The book provides a concise introduction to wind turbine design technologies, foundation structures, construction techniques, operation and maintenance of offshore wind turbines, as well as relevant standards and certification processes. This book can serve as a reference for undergraduate students in engineering disciplines, as well as for graduate students studying offshore wind turbines. It can also be used as a reference by wind power engineering technicians, and certain sections may serve as popular science material for the general public.

The content of the book is structured as follows: Chap. 1 provides an introduction to the development of offshore wind power. Chapter 2 outlines the operational environment of offshore wind turbines. The first two chapters provide readers with background knowledge on offshore wind power and the marine environment. Focusing on offshore wind power and marine environmental conditions, Chap. 3 introduces the design technologies for offshore wind turbines, while Chap. 4 covers the foundation structures of offshore wind turbines. Chapter 5 primarily introduces the construction techniques for offshore wind turbines, while Chap. 6 covers the operation and maintenance of offshore wind turbines. These four chapters present the core components of the book, detailing the system design, construction, operation, and maintenance of offshore wind turbines. To provide readers with a more comprehensive understanding of offshore wind turbines: Chap. 7 discusses the impact of offshore wind power on the marine environment, while Chap. 8 offers a brief introduction to the standards and certification processes for offshore wind turbines.

During the writing of this book, we received valuable assistance from many experts and scholars. Several graduate students from the Wind Power Teaching and Research Office of North China Electric Power University also contributed to the compilation of the manuscript. We would like to extend our sincere thanks to all of them. Given the limitations of the authors, errors and omissions in the book are inevitable, and we kindly ask readers for their constructive criticism and suggestions.

Beijing, China

MingWei Ge
Li Li
Hang Meng
Kai Long

Contents

Chapter 1
Introduction

1.1 The Evolution and Present Condition of Global Offshore Wind Energy

Offshore wind resources are plentiful, situated near load centers, and do not necessitate land use, rendering offshore wind power a vital avenue for wind energy advancement. Since the inaugural offshore wind farm commenced operations in Vindeby, Denmark, in 1991, the offshore wind power sector has evolved swiftly, marked by ongoing innovations in offshore wind turbine technology. In 2009, offshore wind energy constituted merely 1% of the total global installed capacity for wind power. By the conclusion of 2020, this percentage had risen to almost 45%, and the output capacity of individual wind turbines had markedly increased. In 2000, the average capacity of newly erected onshore wind turbines globally was approximately 0.75–1 MW, however by 2020, it had increased to 2.9 MW. In 2000, the mean capacity of newly installed offshore wind turbines was 1 MW, and by 2020, it surpassed 6 MW.

The Global Wind Energy Council (GWEC) anticipates a compound annual growth rate of 31.5% for offshore wind generation from 2021 to 2025. By 2025, the newly installed wind power capacity may quadruple the 6.1GW documented in 2020, with more than 70 GW contributed to the worldwide offshore wind power sector. Fueled by the expansion in emerging markets and the expedited global energy transition, offshore wind power will assume a progressively significant part in the overall expansion of the worldwide wind power market [1].

1.1.1 Examination of Offshore Wind Resources

Offshore wind resources possess distinct advantages to onshore wind, notably greater and more consistent wind speeds. The primary advantages of offshore wind resources can be encapsulated as follows:

M. Ge et al., *Offshore Wind Turbine Technology*,
https://doi.org/10.1007/978-981-96-2175-0_1

Plenty of Offshore Wind Resources: Operational statistics from current offshore wind farms indicate that wind speeds at offshore locations 10 km from the coast are generally approximately 25% greater than those onshore along the coastline.

Reduced Turbulence Intensity: Offshore wind conditions typically exhibit less turbulence intensity and more consistent prevailing wind directions, which can considerably mitigate the fatigue load on wind turbines, hence prolonging their operating lifespan.

The sea surface has reduced roughness relative to land, leading to diminished wind shear effects. This facilitates the design of wind turbines with abbreviated towers, hence diminishing expenses.

Globally, the U.S. National Renewable Energy Laboratory (NREL) employed wind scatter data from the QuikSCAT satellite, in conjunction with reanalysis data from the U.S. National Center for Atmospheric Research (NCAR) and the U.S. National Centers for Environmental Prediction (NCEP), along with data from offshore observation stations, to model the distribution of wind resources at 50 m above sea level. The investigation indicates that wind speeds are minimal near the equator, specifically within the doldrums. Conversely, wind velocities are elevated in proximity to the trade wind belts surrounding the Tropics of Cancer and Capricorn. At approximately 30° latitude in both hemispheres, wind velocities are comparatively diminished within the subtropical high-pressure zones. Wind velocities markedly escalate in the mid-latitude westerlies, particularly in the North Sea area of Europe, where prevailing westerly winds predominate. The Roaring Forties, situated between 40° and 60° latitude in the Southern Hemisphere, are characterized by persistent and vigorous westerly winds throughout the year. Wind velocities are notably elevated in polar locations, characterized by the dominance of polar easterlies.

The allocation of offshore wind resources is closely aligned with the patterns of global pressure and wind belts, as observed from the standpoint of global atmospheric circulation. In certain maritime regions, wind resources demonstrate local variability influenced by factors such as regional climate (e.g., monsoons and land-sea breezes) and terrain (e.g., funneling effects, cape effects, and coastal effects). The Taiwan Strait in China frequently experiences intense winds from the northeast or southwest as a result of funneling effects. The Cape of Good Hope in South Africa frequently experiences elevated wind speeds due to the convergence of airflows resulting from the land extending into the sea, referred to as the cape effect.

The Rutherford Appleton Laboratory in the UK and the RISO Laboratory in Denmark utilized NCAR/NCEP reanalysis data and European marine observation data to assess the offshore wind resource distribution at 50 m above sea level throughout Europe employing the WASP model. The findings indicate that of the three principal seas in Europe—the North Sea, the Baltic Sea, and the Mediterranean—the North Sea possesses the most abundant wind resources, with average annual wind speeds surpassing 9 m/s in the majority of regions. Wind speeds escalate further westward, with certain regions along the UK west coast seeing average wind speeds above 10 m/s. The Baltic Sea possesses substantial wind resources, with average wind velocities of 8 m/s. In comparison, the Mediterranean Sea possesses comparatively inferior wind resources, with the majority of regions exhibiting average wind speeds of approximately 7 m/s.

NREL performed an comprehensive simulation and analysis of the offshore wind resource distribution at 90 m above sea level for the United States, utilizing NCAR/NCEP data, observational station observations, GIS data, and satellite data. The research indicated that the United States had substantial offshore wind resources, with peak wind velocities along the northern East Coast and central West Coast, where they may surpass 9.5 m/s. Wind velocities in the Great Lakes region, the central East Coast, and the Texas shoreline adjacent to the Gulf of Mexico typically range from 8 to 9 m/s.

The National Climate Center of China modeled the distribution of wind resources at an elevation of 70 m above sea level for the country's offshore areas. The findings indicate that China's offshore regions possess substantial wind resources. Of the four principal seas (Bohai Sea, Yellow Sea, East China Sea, and South China Sea), the East China Sea possesses the most abundant wind resources, succeeded by the Bohai Sea, South China Sea, and Yellow Sea. The Taiwan Strait possesses the highest wind energy density among China's offshore regions, with average wind speeds surpassing 8.5 m/s and certain spots exceeding 9 m/s. Mean wind velocities surpass 8 m/s throughout the beaches of Zhejiang and eastern Guangdong, as well as in Liaodong Bay of the Bohai Sea. In the majority of regions, wind speeds exceed 7.5 m/s, but the northern Beibu Gulf and the central Yellow Sea exhibit comparatively inferior wind resources, with average wind speeds of 7 m/s. The nearshore region within the 5 to 25-m depth contour and at 50 m above sea level in China possesses an estimated wind power installation capacity of over 200 GW, signifying considerable offshore wind development potential. China's extensive offshore wind resource reserves create advantageous conditions for the large-scale development of offshore wind power.

1.1.2 The Advancement of Global Offshore Wind Energy

Europe and Asia predominate in the global offshore wind energy sector. The 2021 Global Wind Energy Report published by GWEC on March 25, 2021, indicates that by the conclusion of 2020, the worldwide installed capacity of offshore wind power totaled 35.293 GW, with an increment of 6.068 GW added in 2020 alone. The predominant offshore wind farms are situated in the North Sea, Baltic Sea, Irish Sea, and the English Channel, while the others are dispersed throughout China, Japan, South Korea, and the United States. At the conclusion of 2020, the United Kingdom possessed the highest installed offshore wind capacity, totaling approximately 10.206 GW, succeeded by China with 9.996 GW, Germany with 7.728 GW, and Belgium with 2.262 GW. Denmark, the Netherlands, and Sweden ranked subsequently.

Offshore wind projects in China are predominantly located in three provinces: Guangdong, Fujian, and Jiangsu. Presently, Chinese offshore wind initiatives are primarily situated in nearshore shallow waters, referred to as intertidal zone wind farms. Projects situated farther offshore in deeper seas remain in the development stage.

Europe's offshore wind development commenced early, advancing through three phases: pilot demonstration, large-scale implementation, and commercial development [2]. In 1991, Denmark constructed the world's inaugural offshore wind farm, Vindeby, including 11 wind turbines, each with a capacity of 450 kW. Between 1991 and 2000, offshore wind power was in the experimental stage. In 2001, Denmark's Middelgrunden offshore wind farm, comprising 20 turbines of 2 MW each and a total installed capacity of 40 MW, became the inaugural large-scale offshore wind farm, signifying the commencement of extensive offshore wind deployment in Europe. Subsequently, Europe experienced consistent yearly growth in installed offshore wind power, with individual turbine capacities surpassing 1 MW. As of 2010, the total installed capacity of offshore wind in Europe amounted to 2946 MW.

As of 2011, the mean capacity of newly constructed offshore wind farms in Europe was approximately 200 MW, with the average turbine capacity at 3.6 MW, an average distance from shore of 23.4 km, and water depths of 22.8 m. The offshore wind sector in Europe has commenced a phase of commercial development, characterized by a shift towards larger scales, deeper waters, and more distant offshore sites. In 2012, Belgium's Thornton Bank 2 offshore wind farm included turbines with individual capacity of 6 megawatts. In 2013, the London Array, situated off the UK coast, was established as the world's largest offshore wind farm, boasting a total installed capacity of 630 MW. In 2017, the United Kingdom inaugurated the world's inaugural floating offshore wind farm, Hywind Scotland.

In 2019, global offshore wind capacity rose by an extra 5.2 GW, establishing a record for the greatest annual increment. Globally, 23 offshore wind projects were under construction, totaling a capacity of 7 GW. Notwithstanding the repercussions of the COVID-19 epidemic in 2020, the offshore wind sector experienced expansion, adding over 6 GW of new capacity globally. China dominated global offshore wind capacity, installing nearly 3 GW in 2020 alone.

Currently, China has emerged as the largest offshore wind market globally. Since the installation of its inaugural trial turbine in Bohai Bay in 2007, China's offshore wind sector has expanded swiftly and currently spearheads worldwide offshore wind advancement. As of the end of 2016, China's total installed offshore wind capacity reached 1.6 GW, exceeding Denmark's 1.27 GW, thereby establishing it as the third-largest offshore wind market worldwide. In 2020, China installed 787 offshore wind turbines, contributing 3.845 GW of capacity, thereby establishing itself as the global leader in new offshore wind installations. By the conclusion of 2020, China's total offshore wind capacity surpassed 10 GW, positioning it second worldwide, trailing only the United Kingdom. Figure 1.1 illustrates the global offshore wind capacity expansions from 2016 to 2020.

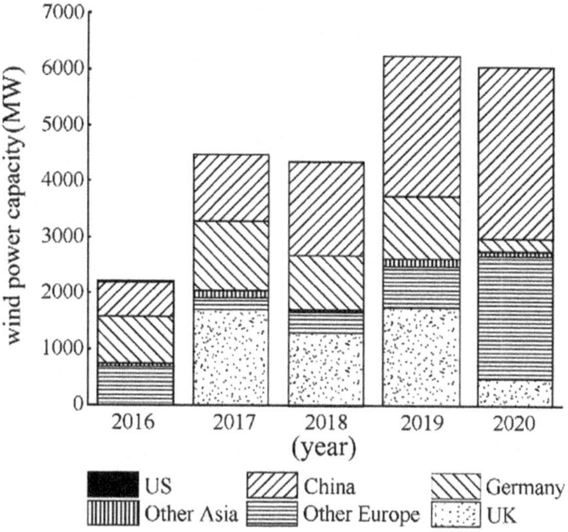

Fig. 1.1 Global offshore wind capacity additions from 2016 to 2020. *Data source* GWEC, 2021 Global Wind Energy Report

1.1.3 Significant Global Offshore Wind Initiatives

The advancement of offshore wind necessitates substantial investment and technological proficiency, resulting in a cost per kilowatt-hour that is typically considerably greater than that of conventional onshore wind. With the swift progression of the offshore wind industry, various large-scale wind developers and equipment manufacturers are diligently investigating creative solutions for offshore wind development, resulting in the proliferation of numerous high-quality and representative offshore wind projects. The following are significant offshore wind projects worldwide:

1. **Hornsea I Offshore Wind Farm**

The Hornsea I offshore wind farm, developed by Ørsted, a prominent entity in offshore wind energy, possesses an installed capacity of 1.2 GW and comprises 174 Siemens Gamesa 7 MW wind turbines. On January 30, 2020, the project achieved complete grid connectivity. Situated in the North Sea, approximately 120 km from the Yorkshire coast, UK, and encompassing 407 km^2, Hornsea I is presently the largest offshore wind farm globally. It was the inaugural facility to produce in excess of 1000 MW of power, providing clean energy to nearly one million households in the UK.

2. **Walney Extension Offshore Wind Farm**

The Walney Extension offshore wind farm, created by Ørsted, is located 19 km from Walney Island, England, and spans roughly 145 km^2. It commenced operations in September 2018. The overall installed capacity is 659 MW, consisting of 87 turbines: 47 MHI Vestas 8 MW turbines with a rotor diameter of 195 m and 40 Siemens Gamesa

7 MW turbines with a rotor diameter of 188 m. This wind farm generates renewable energy sufficient to power around 600,000 residences.

3. **London Array Offshore Wind Farm**

The London Array offshore wind farm is situated 20 km from the coast of Kent, UK, and has 175 Siemens Gamesa 3.6 MW turbines, yielding a total installed capacity of 630 MW. The project, with a total investment of £1.5 billion (about ¥14.487 billion), was the largest offshore wind farm globally until 2018.

4. **Gemini Offshore Wind Farm**

The Gemini offshore wind farm is situated in the North Sea, approximately 85 km from the coast of Groningen, Netherlands, and has an installed capacity of 600 MW. Commissioned in 2017, the wind farm comprises 150 Siemens Gamesa 4 MW turbines, generating sufficient energy to supply around 785,000 residences.

5. **Beatrice Offshore Wind Farm**

The Beatrice offshore wind farm, situated around 13 km from the north coast of Wick, Scotland, commenced operations in July 2019, several months prior to Hornsea I. It possesses an installed capacity of 588 MW and comprises 84 Siemens Gamesa 7 MW turbines. Beatrice provides clean, renewable energy to more than 450,000 households and is projected to decrease carbon emissions by almost 8 million tons over its 25-year lifespan.

6. **Hywind Scotland Floating Offshore Wind Farm**

Commissioned in 2017, Hywind Scotland represents the inaugural floating offshore wind farm globally. The project is situated 25 km east of Peterhead, Scotland, in the North Sea, encompassing an area of 4 km^2 with water depths between 95 and 120 m. The wind farm possesses an installed capacity of 30 MW and comprises five floating wind turbines, each featuring a rotor diameter of 154 m and an overall height of 253 m. The turbines are mounted on spar-type foundations, secured to the seabed by three suction anchors and mooring chains. The floating foundation has a mass of roughly 3200 tons, comprising over 1000 tons of high-density concrete. The Hywind Scotland project, which has a construction cost of £1.9 billion, is jointly run by Norway's Equinor and the UAE's Masdar. The effective operation illustrates the viability of floating offshore wind farms.

1.2 Progress and Present Condition of Offshore Wind Energy in China

China has inherent advantages for offshore wind power development, featuring plentiful offshore wind resources and a vast coastline of 18,000 km. The accessible maritime zone for offshore wind energy development is roughly 300,000 km^2. The fourth national wind resource assessment conducted by the China Meteorological Administration (2008–2010) indicates that the offshore region with water depths ranging from 5 to 25 m and an elevation of 50 m above sea level possesses a potential wind power capacity of around 200 GW.

Despite commencing the development of offshore wind power somewhat late, China's progress has been exceptionally rapid. In 2007, the inaugural offshore wind turbine prototype was deployed in Bohai Bay, and by 2010, China's first large-scale offshore wind farm, the Shanghai Donghai Bridge Wind Farm, was finalized. By 2017, China had constructed 17 offshore wind projects. By the conclusion of 2020, China's cumulative installed offshore wind power capacity surpassed 10 GW for the inaugural time. In 2020, the newly installed capacity exceeded 3845 GW, representing 50.45% of the global offshore wind capacity added that year, thereby establishing China as the global leader for three consecutive years [3]. In September 2020, during the 75th United Nations General Assembly, China declared its "30–60" objective, pledging to peak carbon emissions by 2030 and attain carbon neutrality by 2060. This signifies the commencement of a new era of accelerated expansion in China's offshore wind energy industry.

1.2.1 Examination of Offshore Wind and Wave Conditions in China

The current exploitation of offshore wind resources in China is primarily focused in coastal regions. The coastline region of China, situated on the western boundary of the Pacific Ocean, encompasses tropical, subtropical, and temperate zones. It extends roughly 32° in longitude from east to west and 44° in latitude from north to south. The region undergoes substantial seasonal monsoon fluctuations, leading to marked climate disparities [4]. South of 33° N, prevailing northeasterly winds exhibit average annual wind speeds above 8 m/s in offshore regions from central Zhejiang to eastern Guangdong. Wind velocities in the central Taiwan Strait exceed 9 m/s. This phenomenon is ascribed to the dominant northeasterly winds during winter and southwesterly winds in summer, with Taiwan's terrain conforming to the prevailing wind patterns. The resultant funnel effect greatly enhances the airflow in this area. North of 33° N, the dominant wind direction is more variable, with wind speeds between 6 and 7 m/s. Generally, China's coastal regions can be categorized into four areas according to geographical location: the Bohai Sea and Yellow Sea, characterized by a warm temperate monsoon climate; the East China Sea, which is part of the subtropical monsoon climate; and the South China Sea, predominantly within the tropical monsoon climate zone [5].

1. **The Bohai Sea**

The Bohai Sea is a largely contained inland sea situated at the northern extremity of the eastern Chinese mainland. It extends 55 km from north to south and approximately

236 km from east to west, encompassing a total marine area of 77,284 km^2 and a shoreline measuring 2668 km. The Bohai Sea possesses an average depth of 18 m, but its maximum depth attains 85 m. With over fifty percent of the marine expanse situated at depths less than 20 m, it is the most shallow inland sea in China. The mean annual wind power density in the Bohai Sea varies between 200 and 500 W/m^2, with total hours of wind speeds exceeding 6 m/s totaling 4000 to 6000 h per year. The mean wind velocity in this area ranges from 6 to 9 m/s [6].

The wind patterns in the Bohai Sea demonstrate considerable seasonal fluctuations. Northerly winds dominate in winter, whilst southerly winds rule in summer, with spring and autumn acting as transitional seasons. The average annual wind speed along the coast typically ranges from 5 to 6 m/s, but offshore locations exhibit an average annual wind speed of approximately 7 m/s. Winds rated as force 6 or higher occur on about 50–60 days year, predominantly in winter and spring, with northwesterly or northeasterly winds being the most prevalent.

Wave action in the Bohai Sea is predominantly wind-induced and exhibits marked seasonality. From October to April, northerly swells dominate, and from June to September, southerly swells rule. Wave activity peaks in winter, with average heights between 0.8 and 0.9 m. In summer and autumn, waves generated by typhoons can occasionally surpass 6 m. The most intense wave conditions occur in the Bohai Strait and central Bohai Sea, whilst the Liaodong and Bohai Bays exhibit comparatively weaker waves.

2. **The Yellow Sea**

The Yellow Sea, a marginal sea in the western Pacific, is situated between the Chinese mainland and the Korean Peninsula. It is a semi-enclosed, shallow sea, aligned approximately north–south. The Yellow Sea spans around 380,000 km^2 and has an average depth of about 44 m, including a gently sloping seafloor. The yearly average wind power density in this region varies from 250 to 600 W/m^2, with total hours of wind speeds surpassing 6 m/s amounting to 4000 to 6500 h annually. The mean wind velocity ranges from 7 to 9.5 m/s [6].

The Yellow Sea is significantly affected by monsoons, leading to frigid, arid winters and balmy, humid summers. The wind direction in the Yellow Sea exhibits considerable seasonal variation. From October to March, the northern region is predominantly influenced by northwesterly winds, exhibiting average wind speeds of 6 to 7 m/s. In the southern region, prevailing northerly winds exhibit average speeds of 8 to 9 m/s. April signifies the transitional phase for monsoons, resulting in erratic wind patterns. In May, the southerly monsoon commences, and from June to August, southerly to southeasterly winds prevail, with average wind speeds ranging from 5 to 6 m/s. Winds of force 6 or greater (10.8–13.8 m/s) are present year-round, with peak intensity in winter and highest frequency in spring.

The wave conditions in the northern and southern Yellow Sea exhibit significant differences. The northern section is mostly influenced by wind waves, whilst the southern region encounters more swell waves. From September to April, northwesterly or northerly waves predominate in the north, while northerly waves are predominant in the south. Between June and August, the northern region predominantly experiences southeasterly or southerly waves, whilst the southern region is primarily influenced by southerly waves. Wave activity peaks throughout autumn and winter, particularly in significant wave zones adjacent to Chengshantou and Jeju Island. Wind-generated waves are diminished in size during the spring and summer seasons. Swell waves are generally more elevated in summer and fall compared to winter, with wave heights typically varying from 0.1 to 1.2 m.

3. **The East China Sea**

The East China sea, the most expansive continental shelf water in China, spans approximately 1296 km from northeast to southwest and 740 km from east to west, with a total area of around 770,000 km^2. The continental shelf averages a depth of 72 m, although the whole average depth of the ocean is 349 m, with the deepest point over 2700 m deep. The yearly average wind power density in the East China Sea varies from 500 to 1500 W/m^2, with average wind speeds ranging from 7 to 11 m/s. The total hours of wind speeds exceeding 6 m/s range from 5000 to 7000 h per year [6].

The wind speed and direction in the East China Sea are intricately linked to geographical positioning and seasonal variations. In winter, northerly winds dominate the northern maritime region, whilst northeasterly winds prevail in the southern area, with average wind speeds ranging from 9 to 10 m/s. During spring, northern winds are prevalent in the northern sea, while northeastern winds dominate the south, exhibiting average wind speeds of 5.5–7.5 m/s. During the summer, prevailing southerly winds characterize the whole East China Sea, exhibiting average wind speeds of 5–6.5 m/s. During fall, northeasterly winds dominate over the sea, with average wind speeds of 7–9 m/s.

The monsoon also affects wave conditions in the East China Sea. During spring, waves predominantly originate from the northeast, with average heights varying between 0.8 and 1.4 m. In July, waves predominantly originate from the south, with typical heights of 1 to 1.4 m. During fall, northeasterly waves prevail, with average heights ranging from 1.2 to 2.2 m. During winter, northeasterly waves dominate, with average wave heights ranging from 1.6 to 2.4 m.

4. **The South China Sea**

The South China Sea is the most extensive and profound marginal sea adjacent to China's coastline. Spanning over 3,500,000 km^2, it possesses an average depth of 1212 m, with the maximum depth reaching 5559 m. The yearly average wind power density varies between 500 and 1500 W/m^2, whereas the total hours of wind speeds exceeding 6 m/s range from 5000 to 7000 h per year. The mean wind velocity ranges from 7 to 11 m/s [6].

During winter, the majority of the South China Sea encounters northeasterly winds, with average velocities ranging from 6 to 11 m/s. In spring, east-northeasterly winds dominate the majority of regions, with the Beibu Gulf (Gulf of Tonkin) average wind speeds of 5.5–7 m/s, central maritime areas averaging 4–5.5 m/s, and low-latitude zones experiencing wind speeds below 4 m/s. During summer, southwesterly monsoons prevail, with average wind speeds between 5 and 7.5 m/s. The incidence of severe winds peaks in the northern region of the South China Sea during winter, frequently surpassing 20%, and progressively diminishes southward.

Wave patterns in the South China Sea exhibit seasonal variation. During spring, the northern region is characterized by eastward waves, exhibiting average heights of 1 to 1.4 m, whilst the middle and southern regions are predominantly influenced by northeastward waves, with average heights ranging from 0.8 to 1 m. During summer, typical wave heights in the central and northern parts fluctuate between 1.2 and 1.6 m, although in low-latitude zones, waves do not exceed 0.6 m. In autumn, the northern portion receives northeastward waves ranging from 1.8 to 2.4 m in height, whilst the central area is characterized by northeastward waves of 1.4 to 1.8 m. In the southern section, northward waves dominate, with heights between 0.6 and 1.4 m. During winter, the majority of the South China Sea exhibits average wave heights surpassing 1.6 m.

1.2.2 Advancement of Offshore Wind Energy in China

The eastern coastal regions of China are economically advanced, characterized by substantial electricity demand and a robust power grid, while also featuring superior offshore wind resources, rendering them very conducive to offshore wind power development. In 2007, China deployed a 1.5 MW experimental turbine on an oil rig in Bohai Bay, signifying the inception of offshore wind power production. In 2009 and 2010, China constructed a 32.5 MW experimental wind farm in the intertidal zone of Rudong, Jiangsu Province, with 16 trial turbines. This indicated that China has essentially perfected the technology necessary for the production of large-scale wind turbines, capable of generating turbines over 2 MW, tailored for offshore wind environments. In 2010, China finalized the construction of the 102 MW Donghai Bridge offshore wind farm in Shanghai, achieving full grid connectivity and establishing itself as the nation's inaugural large-scale offshore wind farm. In that year, guided by national energy development planning, the National Development and Reform Commission (NDRC) released the "12th Five-Year Plan for Renewable Energy Development," which sought to expedite wind power development by utilizing the plentiful wind resources and extensive electricity markets along the coast to foster the consistent and proactive progression of offshore wind power.

As of January 2016, China had commissioned around 750 MW of offshore wind power, featuring significant projects such as the Shanghai Donghai Bridge Offshore Wind Demonstration Project (102 MW), Jiangsu Rudong Intertidal Experimental Wind Farm (32.5 MW), Jiangsu Rudong Intertidal Demonstration Wind Farm (200 MW), Jiangsu Xiangshui Nearshore Wind Farm (200 MW), and Jiangsu Rudong Nearshore Wind Farm (150 MW).

In November 2016, the National Energy Administration published the "13th Five-Year Plan for Wind Power Development," establishing a target for cumulative grid-connected wind power capacity to attain a minimum of 210 GW by 2020, with offshore wind capacity surpassing 5 GW. On May 4, 2017, the NDRC and the National Energy Administration collaboratively released the "13th Five-Year Plan for National Marine Economic Development (Public Version)," promoting region-specific and strategically organized offshore wind industry advancement. It promoted the establishment of offshore wind farms in deeper waters and remote places, modification of grid connection policies for wind energy, and enhancement of technical standards for the offshore wind sector and maritime rules.

As of the end of 2019, China had 13 offshore wind projects in progress, representing 56.5% of the global total. China's offshore wind projects under construction possess an installed capacity of 3.7 GW, accounting for 52.9% of the global total. In 2020, China installed 3.845 GW of new offshore wind capacity, representing 50.45% of the global total new capacity, thereby maintaining its leadership in the international market. By the conclusion of 2020, China's cumulative offshore wind capacity surpassed 10 GW, overtaking Germany to establish itself as the world's second-largest offshore wind market, trailing only the United Kingdom.

On September 22, 2020, during the 75th United Nations General Assembly, China declared its "30–60" climate objectives, pledging to reach peak carbon emissions by 2030 and attain carbon neutrality by 2060. China has established a goal to attain a total installed capacity of 1200 GW of wind and solar energy by 2030. In response to this national objective, local governments vigorously advanced the development of offshore wind energy. The Shandong Provincial Development and Reform Commission published the "Shandong Province Mid- to Long-Term Development Plan for New Energy and Renewable Energy (2016–2030)," commencing offshore wind development. The "Strategic Emerging Industry Cluster Development Plan for New Energy (2021–2025)" of Guangdong Province seeks to attain grid parity for offshore wind energy by 2025, targeting a total built offshore wind capacity of 15 GW by that year. On February 10, 2021, the Zhejiang Provincial Development and Reform Commission released the draft "14th Five-Year Energy Development Plan," targeting a total installed wind power capacity of 6.3 GW by 2025, with 5 GW derived from offshore wind.

1.2.3 Exemplary Offshore Wind Power Initiatives in China

1. **Offshore Wind Demonstration Project at Shanghai Donghai Bridge**

The Shanghai Donghai Bridge Wind Farm is China's inaugural large-scale offshore wind demonstration project and Asia's first offshore wind farm. The wind farm, situated 1000 m from either side of the Donghai Bridge linking Lingang New City and Yangshan Deep-Water Port in Shanghai, boasts an installed capacity of 102 MW and a total investment of ¥2.28 billion. The project comprises 34 Sinovel 3 MW offshore wind turbines. The turbines are organized into four units, linked by four 35 kV subsea cables to an onshore 110 kV step-up substation, which incorporates the power into the Shanghai city grid. In 2011, the Donghai Bridge Wind Farm produced

225.84 GWh of energy, achieving an annual full-load equivalent operational duration of 2214 h. The project diminishes normal coal usage by roughly 86,000 tons annually and reduces carbon dioxide emissions by around 230,000 tons each year [7].

The project represented multiple milestones in China's wind farm building history: it was the inaugural deployment of domestically built 3 MW offshore wind turbines, indicating that China's large-capacity wind turbine production has ascended to the forefront of world leaders. This marked the inaugural application of the integrated offshore wind turbine hoisting system, considerably reducing the offshore construction duration. The project achieved a record by erecting 10 turbines on the installation vessel and lifting 8 turbines at sea inside a month. This project was the inaugural implementation of a high-pile cap foundation design globally, successfully addressing the technical issues associated with bearing capacity, uplift resistance, and horizontal displacement for tall wind turbines.

The successful establishment of the Donghai Bridge Offshore Wind Farm expedited the advancement of China's comprehensive offshore wind industry chain, encompassing wind farm design, equipment production, offshore engineering tools, and construction methodologies.

2. **Demonstration Project of the Jiangsu Rudong Offshore Wind Farm**

The Jiangsu Rudong 150 MW Wind Farm is the inaugural intertidal zone wind farm in China, situated in the intertidal seas off Rudong County, Jiangsu Province. The monsoons dictate that prevailing winds originate from the south in summer and from the north in winter, resulting in a relatively scattered yearly wind direction distribution. The prevailing wind direction is southeast, with an annual average wind speed of 7.22 m/s at an elevation of 90 m, and a wind power density of 381 W/m^2, suggesting considerable potential for wind resource development [8].

The wind farm possesses an installed capacity of 149.3 MW, comprising 17 Sinovel 3 MW turbines, 21 Siemens 2.3 MW turbines, and 20 Goldwind 2.5 MW turbines. Construction commenced in June 2011, and the complete wind farm was integrated into the grid and fully operational by April 2013.

3. **Jiangsu Xiangshui Offshore Wind Farm Demonstration Initiative**

The Jiangsu Xiangshui Offshore Wind Farm commenced the installation of its initial wind turbines in August 2015, and by December 2016, all turbines were integrated into the grid. At the time, it was the largest single-unit offshore wind farm in China constructed in a single phase. The wind farm is situated in the offshore region near Guandong Salt Field and Sanxu Salt Field in Xiangshui County, Jiangsu Province, roughly 10 km from the coastline. It spans 13.4 km down the coastline, encompassing a marine area of 34.7 km^2, with water depths varying from 8 to 12 m.

The wind farm has a total installed capacity of 202 MW, with 37 turbines rated at 4 MW each and 18 turbines rated at 3 MW each. Upon being operational, the wind farm is anticipated to produce around 506.69 GWh of power each year. This project will conserve roughly 182,700 tons of standard coal yearly, cut CO_2 emissions by 73,800 tons, and diminish ash residue by 43,500 tons, based on a coal-fired power consumption rate of 320 g of standard coal per kWh.

1.3 Offshore Wind Turbine Technology

1.3.1 Present Condition of Offshore Wind Turbine Technology

Offshore wind turbines encounter considerable hurdles regarding design, manufacture, construction, and elevated operational and maintenance expenses. Consequently, these turbines are advancing towards greater capacity and enhanced intelligence, with floating offshore turbines for deeper oceans becoming progressively sophisticated.

The primary technical specifications of offshore wind turbines encompass:

1. **Enhancing Unit Capacity with Blade Lengths Surpassing 100 m**

To enhance wind energy capture and decrease electricity costs, offshore wind turbines are increasing in size. The blades are essential for transforming wind energy into electrical energy, and their lengths significantly influences the turbine's capacity. Progress in airfoil design, aerodynamic and structural engineering, together enhancements in blade materials and production techniques, has facilitated the development of bigger blade designs. The length of offshore wind turbine blades presently surpasses 100 m. In February 2021, the SR210 blade from Sinomatech measured 107 m, whilst the Vestas V236-15.0 MW design features blades of 115.5 m. The MySE16.0-242, an offshore model by MingYang Smart Energy, is equipped with 118-m-long blades, with prototype production anticipated to commence in 2022 and commercial production scheduled for 2024.

2. **Enhanced Intelligence and Automation**

Offshore wind farms necessitate substantial investment and are situated in demanding conditions, presenting access challenges for maintenance. These issues necessitate elevated standards of reliability and sophisticated design in wind turbines. To optimize energy generation while reducing expenses in design, operation, and maintenance, cutting-edge digital intelligence technologies are progressively utilized throughout the whole lifecycle of offshore wind farms. This encompasses resource evaluation, tailored wind farm design, advanced management systems, and intelligent operation platforms that incorporate fault diagnostics, health monitoring, power output forecasting, and operational enhancements. These platforms employ advanced technologies such as big data, cloud storage, and the Internet of Things (IoT) [9].

3. **Improved Adaptability to Marine Environments**

Offshore wind turbines are increasingly designed to accommodate marine conditions, taking into account concerns such as seawater corrosion, elevated humidity, and severe weather. Offshore wind turbines necessitate more refined designs compared to onshore ones, integrating anti-typhoon measures, corrosion resistance, anti-collision structures, and seismic protection. Moreover, sophisticated control systems are utilized to guarantee their reliability and robustness in marine conditions.

4. **Maturing Floating Wind Turbines for Deep Water Applications**

The intricate and hazardous situations in deeper waters render conventional permanent foundations unfeasible due to their substantial size and elevated expense. Floating foundations provide enhanced versatility, coupled with reduced construction and maintenance expenses, rendering them more feasible for deepwater applications [10]. Floating foundations often comprise anchor chains, anchors, and buoyant boxes. The state-owned oil corporation of Norway has effectively deployed the Hywind single-column floating technology on a limited commercial scale. Simultaneously, alternative floating foundation technologies, such the WindFloat semi-submersible platform created by Principle Power Inc. and the dampening pool semi-submersible design by Ideol, are progressing towards commercial feasibility.

5. **Ongoing Enhancement of Technical Standards**

Standards for offshore wind turbines are consistently advancing, based on those developed for onshore turbines. International standards, including IEC 61400-1 (pertaining to floating wind turbine design) and IEC 61400-3 (related to fixed offshore turbine design), constitute essential guidance. As offshore wind technology and the industry progress in China, national requirements are concurrently being enhanced. This encompasses standards for the design, surveying, operation, and maintenance of offshore wind farms, including the Design Standard for Offshore Wind Farms (GB/T 51308-2019) and the Technical Requirements for Offshore Wind Turbine Foundations (GB/T 36569-2018).

Additionally, industry-specific standards pertaining to wind resource assessment, marine hydrological monitoring, and the operation of unmanned offshore substations are facilitating the robust advancement of the offshore wind sector.

This improved regulatory framework facilitates the ongoing development and optimization of offshore wind turbine technology, guaranteeing safer, more efficient, and more economical operations.

1.3.2 Prominent International Large Offshore Wind Turbines

1. **Vestas V174-9.5 MW and V236-15.0 MW Offshore Wind Turbines**

(1) V174-9.5 MW Offshore Wind Turbine

The V174-9.5 MW offshore wind turbine possesses a rated power of 9.5 MW, featuring a rotor diameter of 174 m and blade length of 85 m. The nacelle has dimensions of 21 m in length, 9 m in width, and 9 m in height, with an approximate weight of 390 tons. The hub height is roughly 110 m, while the blade tip attains a height of around 197 m. The swept area measures around 23,779 m^2. The V174-9.5

MW offshore wind turbine was inaugurated in February 2019. Essential technical specifications are presented in Table 1.1.

(2) V236-15.0 MW Offshore Wind Turbine

On February 7, 2021, Vestas launched the V236-15.0 MW offshore wind turbine. This turbine model attains a height of 260 m, features a rotor swept area surpassing 43,000 m^2, and produces an annual energy production of up to 80 GWh. The turbine is engineered for a lifespan exceeding 25 years. The prototype of the V236-15.0 MW offshore wind turbine was deployed in 2022, with mass production anticipated to commence in 2024. The technical specifications of the V236-15.0 MW offshore wind turbine are outlined in Table 1.2.

2. **General Electric (GE) Offshore Wind Turbines: Haliade-X and Haliade 150-6MW (1) Haliade-X Offshore Wind Turbine**

Table 1.1 Technical specifications of the V174-9.5 MW offshore wind turbine

Parameter type	Technical specifications
Rated power	9500/9600 kW
Cut-in wind speed	3 m/s
Cut-out wind speed	25 m/s
Turbine class	IEC IB or IIB, T (suitable for offshore environments)
Standard operating temperature range	− 15 to + 25 °C, with derating between + 25 and + 35 °C
Maximum noise level	112.9 dB(A)
Rotor diameter	174 m
Swept area	23,779 m^2
Aerodynamic braking system	Three blade pitch regulation

Table 1.2 Technical specifications of the V236-15 MW offshore wind turbine

Parameter type	Technical specifications
Rated power	15,000 kW
Cut-in wind speed	3 m/s
Cut-out wind speed	30 m/s
Turbine class	IEC S or S, T (suitable for offshore environments)
Standard operating temperature range	− 10 to + 25 °C, with derating between + 25 and + 45 °C
Maximum noise level	118 dB(A)
Rotor diameter	236 m
Swept area	43,742 m^2
Aerodynamic braking system	Three blade pitch regulation

(1) Haliade-X Offshore Wind Turbine

The Haliade-X series of offshore wind turbines manufactured by General Electric (GE) has versions with capacities of 14 MW, 13 MW, and 12 MW. The Haliade-X turbine features blades that are 107 m long, with a rotor diameter of 220 m. The turbine's tip height attains 248 m, encompassing a swept area of 38,000 m^2. The projected yearly energy output is predicted to be between 63 and 74 GWh. The technical characteristics of the Haliade-X offshore wind turbine are detailed in Table 1.3.

(2) Haliade 150-6 MW Offshore Wind Turbine

The Haliade 150-6 MW offshore wind turbine is engineered for diverse offshore environments. The rotor diameter measures 150 m, with blade extensions of 73.5 m. The technical specifications of the Haliade 150-6 MW are detailed in Table 1.4.

3. **SIEMENS Gamesa's SG 14-222 DD and SG 80-167 DD Offshore Wind Turbines**

(1) SG 14-222 DD Offshore Wind Turbine

On May 19, 2020, Siemens Gamesa unveiled the SG 14-222 DD offshore direct-drive wind turbine. This type possesses a rated capacity of 14 MW, a rotor diameter

Table 1.3 Technical specifications of the Haliade-X offshore wind turbine

Turbine model	Haliade-X 12 MW	Haliade-X 13 MW	Haliade-X 14 MW
Output power (MW)	12	13	14
Rotor diameter (m)	220	220	220
Total height (m)	248	248	248
Frequency (Hz)	50 and 60	50 and 60	50 and 60
Annual energy production (GWh)	$\leq$ 68	$\leq$ 71	$\leq$ 74
Capacity factor (%)	63	60–64%	60–64%
IEC wind turbine class	IB	IC	IC

Table 1.4 Technical specifications of the Haliade 150-6 MW offshore wind turbine

Parameter type	Technical specifications
Rated power	6 MW
Hub height	100 m
Rotor diameter	150 m
Swept area	17,860 m^2
IEC wind class	I-B (IEC 61400-1/IEC 61400-3)
Tower type	Steel tube
Rotor type	Permanent magnet direct drive

Table 1.5 Technical specifications of the SG 14-222 DD offshore wind turbine

Parameter type	Technical specifications
IEC class	I, S
Rated power	14 MW
Rotor diameter	222 m
Blade length	108 m
Swept area	39,000 m^2
Hub height	Specific to location
Power regulation	Pitch regulation, variable speed

Table 1.6 Technical specifications of the SG 8.0-167 DD offshore wind turbine

Parameter type	Technical specifications
Rated power	8.0 MW
iec class	I, S
Rotor diameter	167 m
Blade length	81.4 m
Swept area	21,900 m^2
Hub height	Specific to location
Power regulation	Pitch regulation, variable speed

of 222 m, and a blade length of 108 m, encompassing a swept area of 39,000 m^2. The prototype design of the SG 14-222 DD is set for completion in 2021, with commercial release anticipated in 2024. The technical characteristics of the SG 14-222 DD offshore wind turbine are detailed in Table 1.5.

(2) SG 8.0-167 DD Offshore Wind Turbine

The SG 8.0-167 DD offshore wind turbine is engineered for the climatic conditions of the Asia–Pacific area, accommodating typhoons, seismic activity, and extreme temperature variations. This direct-drive turbine adheres to regional requirements and is ideally suited for offshore wind markets in the Asia–Pacific, including Taiwan. The technical parameters of the SG 8.0-167 DD are detailed in Table 1.6.

1.3.3 Major Large-Scale Offshore Wind Turbines in China

1. **Goldwind's GW 175-8.0 MW and GW 6X MW Offshore Wind Turbines**

(1) GW 175-8.0 MW Offshore Wind Turbine

The GW 175-8.0 MW offshore wind turbine possesses a rated power of 8 MW and a rotor diameter of 175 m. The technical characteristics of the GW 175-8.0 MW are detailed in Table 1.7.

Table 1.7 Technical specifications of the GW 175-8.0 MW offshore wind turbine

Parameter type	Technical Specifications
Rated power	8.0 MW
IEC wind class	IEC S
Cut-in wind speed	3 m/s
Rated wind speed	12 m/s
Cut-out wind speed	25 m/s (25–30 m/s in storm mode)
Design life	25 years
Operating temperature	− 10 to + 40 °C
Survival temperature	− 20 to + 50 °C
Rotor diameter	175 m
Swept area	23,916 m^2
Adaptability	Offshore, Typhoon-resistant

(2) GW 6.X MW Offshore Wind Turbine

The GW 6.X MW offshore wind turbine platform employs direct-drive permanent magnet technology. The technical details are detailed in Table 1.8.

2. **Shanghai Electric's GS 8.0-167 and 625-172 Offshore Wind Turbines for Typhoon Conditions and Low-to-Medium Wind Speeds**

(1) GS 8.0-167 Offshore Wind Turbine

The GS 8.0-167 offshore wind turbine represents the highest capacity offering from Shanghai Electric's direct-drive platform. Engineered for IEC Class I wind zones, the turbine's blades, tower, anemometer, pitch lock, and control system have been enhanced to withstand diverse typhoon conditions. It adheres to many design requirements, including IEC 61400, DNV-GL tropical cyclone turbine certification,

Table 1.8 Technical specifications of the GW 6.X MW offshore wind turbine

Turbine model	GW154/6700	GW164/6450	GW171/6450
Rated power (kW)	6700	6450	6450
IEC wind class	IEC S	IEC S	IEC III B
Cut-in wind speed (m/s)	3	3	3
Rated wind speed (m/s)	12	11.5	10.5
Cut-out wind speed (m/s)	25	25	25
Design mean wind speed (m/s)	10	8.5	7.5
Max 10-min wind speed (m/s)	57	50	37.5
Design life (years)	25	25	25
Rotor diameter (m)	154	164	171
Swept area (m^2)	18,617	21,124	22,960

Table 1.9 Technical specifications of the GS 8.0-167 offshore wind turbine

Parameter type	Technical specifications
Rated power	8.0 MW
Blade length	81.4 m
Cut-in wind speed	3 m/s
Rated wind speed	12–14 m/s
Cut-out wind speed	25 m/s
Design life	25 years
Rotor diameter	167 m
Swept area	21,900 m^2
Wind class	IEC I

and China's GB/T 31519 standard for typhoon-resistant wind turbines. The turbine incorporates full-power variable frequency technology, employs a permanent magnet synchronous motor, and has a hydraulic pitch control system. The blades are produced by a singular molding technique. The technical specifications are detailed in Table 1.9.

(2) GS 6.25-172 Offshore Wind Turbine for Typhoon Conditions and Low-to-Medium Wind Speeds

The technical specifications of the GS 6.25-172 offshore wind turbine designed for typhoon circumstances and low-to-medium wind speeds are detailed in Table 1.10.

3. **MingYang Wind Power MySE 6.0 MW Series Wind Turbines**

The MySE 6.0 MW series wind turbines utilize a semi-direct-drive compact transmission technology. They are explicitly engineered to suit China's offshore wind

Table 1.10 Technical specifications of the GS 6.25-172 offshore wind turbine for typhoon conditions

Parameter type	Technical specifications
Wind class	IEC S
Rotor diameter	172 m
Swept area	23,235 m^2
Power regulation method	Pitch control, variable speed
Cut-out wind speed	25 m/s (25–30 m/s in storm mode)
Rated power	6250 kW
Hub height	110 m (customized for specific projects)
Cut-in wind speed	3 m/s
Cut-out wind speed	25 m/s
Operating temperature range	– 20 to 40 °C
Survival temperature range	– 30 to 50 °C

resources and maritime ecosystem. These turbines provide high power production efficiency, robust reliability, exceptional corrosion resistance and heat dissipation, great typhoon resilience, and minimal operational and maintenance expenses. The fundamental specifications of the MySE 6.0 MW series wind turbines are presented in Table 1.11.

4. **CSIC Haizhuang's H151-5 MW and H152-6.2 MW Offshore Wind Turbines**

(1) H151-5 MW Offshore Wind Turbine

On October 15, 2016, CSIC Haizhuang commenced mass production of its 5 MW offshore wind turbine. The technical specifications are presented in Table 1.12.

(2) H152-6.2 MW Offshore Wind Turbine

In July 2020, CSIC Haizhuang officially launched the H152-6.2 MW offshore wind turbine. The technical specifications are presented in Table 1.13.

Table 1.11 Basic parameters of the MySE 6.0 MW series wind turbines

Turbine model	MySE 5.5-155	MySE 7.25/ 7.0-158	MySE 6.45-180	MySE 8.3-180
Rated power (kW)	5500	7250/7000	6450	8300
Design wind class	IEC IB	IEC IB	IEC S	IEC S
Cut-in wind speed (m/s)	3	3	3	3
Rated wind speed (m/s)	10.1 (static)	11.1 (static)	10.5 (static)	11.3 (static)
Cut-out wind speed (m/s)	30 (3 s average)	30 (3 s average)	30 (3 s average)	30 (3 s average)
Design life (years)	25	25	25	25
Suitable environment	Offshore, typhoon-resistant			

Table 1.12 Technical specifications of CSIC Haizhuang's 5 MW offshore wind turbine

Turbine model	H127-5 MW	H151-5 MW
Rated power (kW)	5000	5000
Rotor diameter (m)	127	151
Cut-in wind speed (m/s)	3.5	3.5
Rated wind speed (m/s)	12.1	10.6
Cut-out wind speed (m/s)	25	25
Design life (years)	25	25
Design availability	≥ 95%	≥ 95%

Table 1.13 Technical specifications of the H152-6.2 MW offshore wind turbine

Parameter type	Technical specification
Rated power	6.2 MW
Applicable wind zone	IEC IB
Cut-in wind speed	3 m/s
Rated wind speed	10.5 m/s
Cut-out wind speed	25 m/s
Design service life	25 years
Rotor diameter	152 m
Hub height	96 m

Table 1.14 Technical specifications of the XE6.X-7.X series offshore wind turbines

Turbine model	XE174-6450	XE185-6500	XE174-7250	XE174-8000
Rated power (kW)	6450	6500	7250	8000
Rotor diameter (m)	174	185	174	174
Cut-in wind speed (m/s)	3	3	3	3
Rated wind speed (m/s)	11	10.5	11.5	12.5
Cut-out wind speed (m/s)	25	25	25	25
Design life (years)	25	25	25	25
Swept area (m^2)	23,778	26,880	23,778	23,778
Applicable wind class	IEC S	IEC S	IEC S	IEC S

5. **Xiangdian Wind Power's XE6.X-7.X Series and XE128-5000 Offshore Wind Turbines**

(1) XE6.X-7.X Series Offshore Wind Turbines

The XE6.X-7.X series offshore wind turbines from Xiangdian Wind Power employ direct-drive permanent magnet generator technology and medium-voltage frequency converter technology. The technical specifications are presented in Table 1.14.

(2) XE128-5000 Offshore Wind Turbine

The XE128-5000, produced by Xiangdian Wind Power, is China's inaugural 5 MW offshore wind turbine. In November 2015, a total of 10 units were deployed in Pinghai Bay, Putian, Fujian Province. The technical characteristics of the XE128-5000 offshore wind turbine are detailed in Table 1.15.

Table 1.15 Technical specifications of the XE128-5000 offshore wind turbine

Parameter type	Technical specifications
Rated power	5 MW
Applicable wind class	IEC IB
Cut-in wind speed	3 m/s
Rated wind speed	11.5 m/s
Cut-out wind speed	25 m/s
Design life	25 years
Rotor diameter	128 m
Swept area	12,773 m^2

References

1. Global Wind Report 2021. Global Wind Energy Council (GWEC), 2021.
2. Zhang B, Lu B. Construction of offshore wind farms in Europe. Shanghai Electr Power, 2007;20(2):7.
3. Liu Z, Yu J, Huang D, et al. Research on the development of offshore wind power. Ship Eng. 2020;42(8):6.
4. Qi H. Climatic zoning of offshore wind resources in China. Qingdao: Ocean University of China; 2015.
5. Zhang J, Fu P, Ling Y. Offshore wind farm construction technology and practice. Beijing: China Water & Power Press; 2013.
6. National Marine Awareness Development Index Research Group. National Marine Awareness Development Index Report. Beijing: Ocean Press; 2017.
7. Lin Y, Li J, Shen D, et al. Characteristics and design of wind turbine foundations for Donghai bridge offshore wind farm. Shanghai Electr Power 2007;20(2):5.
8. Qiu Y. Overview of offshore wind farm development. 2017;11:38–41.
9. Xu G, Ye H, Xie H. Current status and development trends of wind turbine technology. Eng Sci China 2018;20(3):7.
10. Liu C, Xu Y. Analysis of the development prospects of floating offshore wind power in China. China Foreign Energy; 2020.

Chapter 2
Operational Environment of Offshore Wind Turbines

The offshore environment is intricate, and the design and operation of wind turbines are intimately linked to offshore wind conditions, ocean hydrology, and geological factors. In contrast to Europe, China's coastal regions often encounter typhoons, and the hydrological and geological conditions are comparatively more intricate. This chapter will delineate the operational environment of offshore wind turbines, focusing on typhoons, hydrology, and geological variables.

2.1 Typhoons

Typhoons are a classification of intense cyclones that originate over tropical seas. In the western North Pacific, west of the International Date Line, including the South China Sea, a vigorous tropical storm with wind speeds over 32.6 m/s is classified as a typhoon. A typhoon often delivers severe winds and significant rainfall to an area. A typhoon, a kind of cyclone, features low air pressure at its center, causing surface winds to converge from all directions toward the core. In the Northern Hemisphere, typhoon winds rotate in a counterclockwise direction.

The national standard "Tropical Cyclone Classification" (GB/T 19201—2006) classifies tropical cyclones into six categories based on the maximum average wind speed at the cyclone's surface center. Consult Table 2.1 for details [1].

The severe winds and considerable precipitation linked to typhoons during their passage markedly impact the operation of wind turbines. In extreme offshore weather conditions, understanding the origin, structure, trajectory characteristics, intensity, spatial–temporal patterns, and landfall dissipation of typhoons is crucial for the installation of wind turbines.

M. Ge et al., *Offshore Wind Turbine Technology*,
https://doi.org/10.1007/978-981-96-2175-0_2

Table 2.1 Classification of tropical cyclones

Categories of tropical cyclones	Peak mean wind velocity at the lower center (m/s)	Peak wind intensity adjacent to the lower center (scale)
Tropical depression (TD)	10.8–17.1	6–7
Tropical storm (TS)	17.2–24.4	8–9
Severe tropical storm (STS)	24.5–32.6	10–11
Typhoon (TY)	32.7–41.4	12–13
Severe typhoon (STY)	41.5–50.9	14–15
Super typhoon (super TY)	≥ 51	16 and above

2.1.1 *Conditions for Typhoon Formation*

Typhoons occur across vast tropical oceans located between 5° and 25° north and south latitudes. In tropical seas, direct sunlight heats the ocean surface, promoting the evaporation of saltwater into water vapor that ascends into the sky. The air over the tropical ocean expands due to increased temperatures, resulting in decreased density and mass. Furthermore, owing to the weak winds near the equator, this air ascends freely, leading to convective activity. Simultaneously, neighboring cold air infiltrates to refill the rising air, creating an ongoing cycle. This process ultimately saturates the whole air column with warmer, lighter, and less dense air, leading to the development of a tropical depression. Air transitions from areas of high pressure to areas of low pressure, causing neighboring high-pressure air to flow into the low-pressure zone, therefore generating wind. A wind speed of 32.6 m/s or more at the surface is classified as a typhoon.

The development of typhoons necessitates the following conditions:

- An adequately extensive tropical ocean surface is necessary. The temperature and humidity of the lower atmosphere above the tropical ocean are primarily determined by the ocean surface temperature, which must surpass 26.5 °C. This temperature must be maintained at depths of up to 60 m in the ocean.
- Before the formation of a typhoon, a weak tropical vortex must be present. The formation and development of a typhoon require energy sourced from the water vapor above the tropical ocean surface. In a defined tropical vortex, the interior pressure is lower than that of the surrounding environment. As a result, ambient air, saturated with significant water vapor, converges toward the vortex's center and continuously rises within it. The rising humid air causes water vapor to condense, producing significant latent heat that promotes typhoon formation.
- A significant Coriolis force must result from the Earth's rotation. The Earth's rotation affects the formation of cyclonic vortices. The Coriolis force diminishes toward the equator and increases toward the poles, elucidating why typhoons primarily develop over maritime areas located more than 5 degrees from the equator.

- In the vicinity of a weak low-pressure system, the differences in wind direction and speed between high and low-pressure air must be negligible. The vertical air column ascends evenly, promoting heat retention in the upper atmospheric strata, resulting in warming. When a cyclone forms, the atmospheric airflow above the friction layer will move across the isobars, so amplifying the warming effect. In regions north of 20°N, climatic conditions change mostly due to strong upper-level winds that obstruct warming and inhibit typhoon development.

2.1.2 The Structure of a Typhoon

The circulation pattern of a typhoon is virtually axially symmetric, like that of a cylinder. The typhoon region is divided into three components: the eye of the typhoon, the vortex rainband, and the outer wind zone. The typhoon's eye is located at its center, with a diameter of 5–10 km. The central pressure frequently varies between 980 and 950 hPa, with the possibility of declining to a minimum of 870 hPa. Mature typhoons often feature a distinct eye, with the eyewall signifying the area of maximum convection and the highest horizontal wind speeds. As the air in the eyewall rises to higher convective levels, it is constrained by the tropopause, leading to outward divergence in an anticyclonic manner. A small volume of air converges and drops toward the center, forming the eye of the typhoon. The vortex rainband surrounding the eye may display maximum wind velocities exceeding 17 on the Beaufort scale. The external wind zone extends beyond this juncture. The radius of the typhoon's vortex often ranges from 500 to 1000 km [2].

Figure 2.1 depicts the schematic representation of the typhoon structure.

In the low-latitude oceanic regions of the Northern Hemisphere, typhoon vortices rotate counterclockwise and are influenced by the Coriolis force, leading to a pole-ward movement. Simultaneously, easterly winds at low latitudes guide their west-ward trajectory. The typhoon cyclone advances on a northwest course because to the interaction of these factors. Upon attaining higher latitudes, it reaches the wester-lies and, influenced by southwestern winds, modifies its course to advance northeast. Subsequent to the influx of cold air, the typhoon metamorphoses into an extratropical storm; some merge with westerly troughs, while others dissipate over cold oceanic surfaces. This approach illustrates a complete and typical life cycle of a typhoon.

2.1.3 Typhoon Climate Characteristics in China's Coastal Regions

Typhoons mostly originate in the Northwest Pacific, constituting around one-third of global typhoons, and are generally the most potent. China, located on the North-west Pacific coast, is one of the countries most frequently affected by typhoons. The

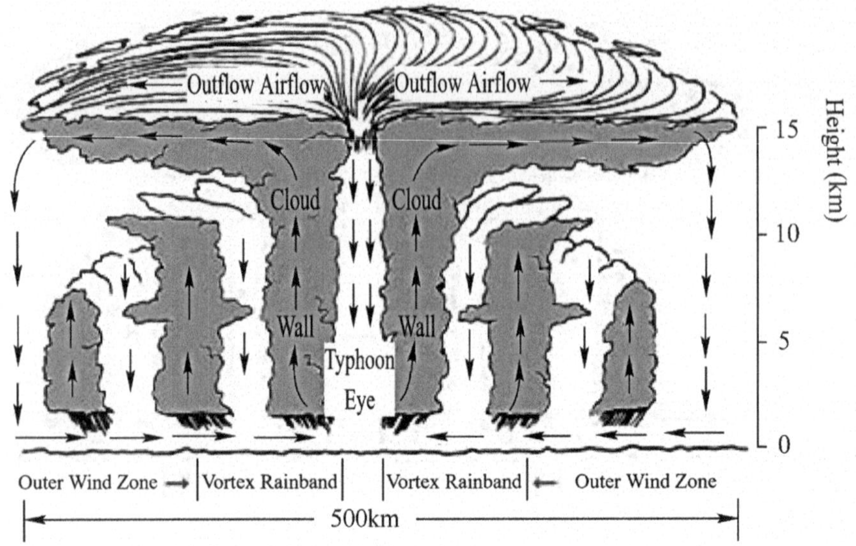

Fig. 2.1 Schematic diagram of typhoon structure. *Source* National Meteorological Information Center

southern coastal regions of China are profoundly impacted by the frequent occurrence of typhoons. Guangdong has an average of three landfalling typhoons year, accounting for 33% of the total annual typhoon landfalls in China. Taiwan accounts for 19%, Hainan for 17%, Fujian for 16%, and Zhejiang for 10% [3].

The "Technical Specification for Assessing the Impact of Tropical Cyclones on Offshore Wind Farms" (GB/T 38957-2020) stipulates that the statistical characteristics of tropical cyclones in the risk assessment area must encompass the annual and monthly average frequency and rates of tropical cyclones across various categories, in addition to the frequency and rates of diverse typhoon trajectories. Data from the CMA's Best Track Data Collection indicate that between 1949 and July 2018, a total of 2299 typhoons formed in the Northwest Pacific area, with 627 making landfall on China's coast. This constitutes 27.3% of the total typhoons produced in the Northwest Pacific. Annually, an average of 32.8 typhoons are formed, with 9 making landfall in China. There is considerable interannual variability in the frequency of landfalls, peaking in 1952 and 1961, with each year recording 15 occurrences. Other significant years include 1967, 1985, 1989, and 1994, each recording 13 landfalls, although 1982 had the fewest with only 4. The maximum yearly landfalls are almost three times the minimum [4].

Figure 2.2 illustrates the interannual variance of typhoon landfalls in China from 1949 to 2019.

In terms of the monthly distribution of typhoon landfalls in China, there are essentially no typhoons that make landfall between January and March. Typhoons make landfall in various months, predominantly from July to September, which

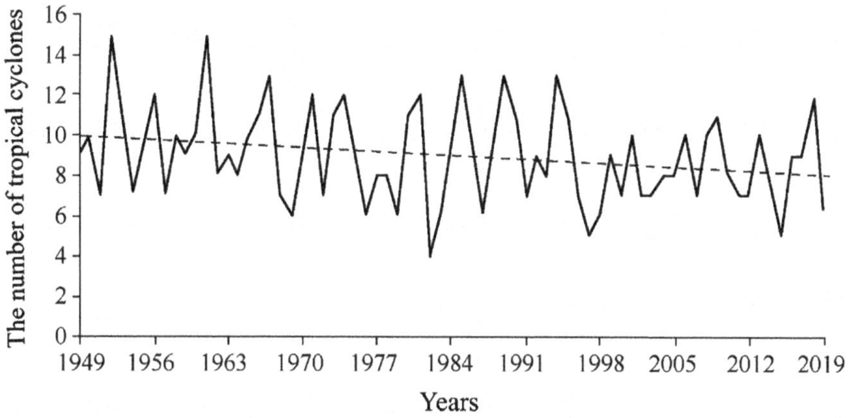

Fig. 2.2 Interannual variation of typhoon landfalls in China from 1949 to 2019

constitutes around 76.3% of the total yearly occurrences. August has the highest frequency of landfalls, accounting for around 28% of the yearly total. In 1997, all five typhoons that struck land occurred in August. Landfalls are few before and after the July to September interval, occurring only twice in April and twice in December. Figure 2.3 illustrates the monthly fluctuations in the incidence of typhoons making landfall in China from 1949 to 2019.

All coastal provinces in China, with the exception of Hebei, Tianjin, and Macau, have encountered direct landfalls of typhoons. The primary landing sites are located in Guangdong, Hainan, and Taiwan, comprising 82% of the overall landfalls. Guangdong possesses the biggest quantity with 221, constituting 35.2% of the whole total. Fujian and Zhejiang occupy the fourth and fifth positions, respectively, although Guangxi has the least landfalls, totaling only 2. Figure 2.4 illustrates the pattern of typhoon landing places from 1949 to 2019.

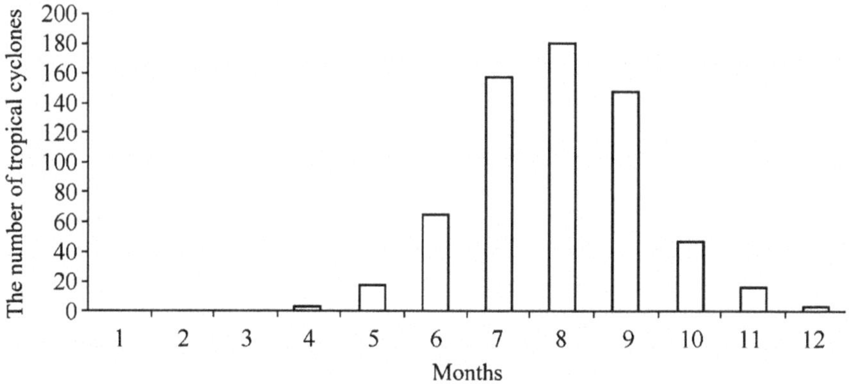

Fig. 2.3 Monthly variation of typhoon landfalls in China from 1949 to 2019

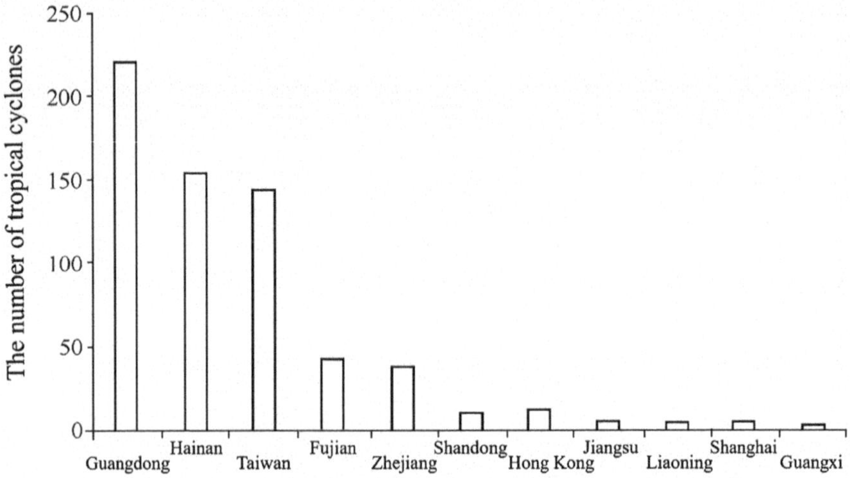

Fig. 2.4 Distribution of typhoon landing points from 1949 to 2019

The trajectory and velocity of a typhoon's movement are dictated by the forces exerted upon it, which can be categorized as internal and external factors. Internal forces arise from the disparities in Coriolis force due to latitude change, resulting in a net force directed northward and westward. An increase in internal force correlates with an expanded range of the typhoon and heightened wind velocities. External forces refer to the impact of the ambient flow field surrounding the typhoon on its vortex. Internal causes generally influence the initial development of the typhoon, however external forces predominantly dictate its path. Consequently, typhoons often traverse from east to west.

The trajectories of typhoons are irregular due to the configuration, strength, location of the subtropical high, and several other characteristics. Despite the varied trajectories of typhoons in the Northwest Pacific, certain patterns emerge in the internal and external factors. The primary movement trajectories can be classified as westward, landing, and parabolic types. Westward Classification. This trajectory extends from the eastern Philippines directly westward, traversing the South China Sea and reaching landfall in Hainan Island or northern Vietnam, predominantly during October and November. Type of Landing. This system progresses northwest, traversing the Taiwan Strait to reach the shores of Guangdong, Fujian, and Zhejiang in China, eventually diminishing into a low-pressure zone. This trajectory exerts the most significant influence on China, predominantly throughout July and August. Parabolic Form. This kind initially progresses northwest but shifts to a northeast trajectory as it nears the eastern coastal regions of China, passing close to Japan without making landfall in China. The trajectory exhibits a parabolic form and predominantly transpires during May to June and September to November [5].

The trajectories of typhoons making landfall on China's coast display seasonal fluctuations, strongly linked to the seasonal modifications of atmospheric circulation, notably the shifts in the position of the Pacific subtropical high.

2.1.4 The Impact of Typhoons on the Design of Wind Turbines

To guarantee the safety and sustained dependable functioning of offshore wind turbines, it is imperative to assess the influence of the maritime environment on their functionality throughout the design process. The negative impacts largely emerge in many areas, including loads, service life, and operation and maintenance. Environmental circumstances are categorized as regular external conditions and extreme external conditions: normal external conditions pertain to prolonged fatigue loads and operational states, whereas extreme external conditions denote possible critical external design situations. Typhoons and tropical cyclones of equivalent or greater strength should be regarded as critical external circumstances in the design of wind turbines and the selection of equipment for wind farms [6, 7].

1. **Extreme wind speed**

Typhoons are low-pressure systems, and due to horizontal pressure differentials, surrounding air converges from all directions into the core of the typhoon. In the Northern Hemisphere, the Coriolis effect causes inflowing air to rotate counterclockwise, with tangential wind speed increasing as it spirals inward toward the typhoon's core. The wall of clouds around the typhoon eye is generally the region of most and most concentrated destructive force, spanning a breadth of 10–20 km. The most intense winds are often located in the right front quadrant of the typhoon's trajectory. As the typhoon's center nears, wind velocities at all altitudes progressively escalate. Upon the passage of the typhoon's eye, there is an abrupt and considerable reduction in wind velocity, succeeded by a swift escalation 10 to 30 min later (contingent upon the eye's dimensions and translational speed), after which the wind speed progressively diminishes as the typhoon recedes. The severity of the northern winds often peaks immediately prior to the arrival of the typhoon's eye.

In the design of wind turbines, it is essential to account for the maximum wind speed that is expected to occur once every 50 years. The principal data sources include of meteorological stations situated along the southeast coast and wind measurement towers, which furnish observational data on tropical storms. A dataset of maximum wind speeds for tropical storms over a 50-year period is compiled using the gathered observational data. The maximum wind speed for a 50-year event is determined using the Poisson-Gumbel joint probability distribution. The maximum wind speed at an elevation of 70 m is projected based on the vertical wind shear data obtained from coastal wind monitoring towers. Coastal wind farms assess and compute the maximum wind speed for a 50-year event by conducting pertinent assessments of meteorological stations and wind measuring towers.

Typhoon occurrences may transpire many times year, or they may not manifest at all. The Poisson-Gumbel joint probability distribution model is appropriate for highly random phenomena, such as hurricanes or typhoons. It is assumed that the frequency of typhoon impacts p_k follows a Poisson distribution, denoted as

$$p_k = e^{-\lambda}\frac{\lambda^k}{k!}, k = 0, 1, 2, \ldots \tag{2.1}$$

In the equation, $\lambda = \frac{N}{n}$, where N represents the total number of typhoon impacts and n denotes the total number of years.

Assuming the wind speed under typhoon conditions follows a Gumbel distribution, denoted as

$$G(x) = \exp\{-\exp[-\alpha(x-\delta)]\} \tag{2.2}$$

The cumulative distribution function of the Poisson-Gumbel compound extreme value distribution is given by

$$F(x) = \sum_{0}^{k} p_k[G(x)]^k = \exp\{-\lambda[1-G(x)]\} = P \tag{2.3}$$

By rearranging the above formula, the maximum wind speed corresponding to a probability P is given by

$$v_p = \delta + \frac{-\ln\{-\ln[1+\frac{1}{\lambda}\ln(1-\frac{1}{T})]\}}{\alpha} \tag{2.4}$$

In the formula, α and δ are obtained from the Gumbel distribution, where $\alpha = 1.28255/\sigma, \delta = \overline{x} - 0.57722/\alpha$, Here, $\overline{x}$ is the mean of the sample sequence, and σ is the standard deviation of the sample sequence.

This approach basically guarantees that the robust wind samples derive from distinct typhoon occurrences. To align the sample sequence with the Poisson distribution, a wind speed threshold must be determined; data beyond this barrier will be incorporated into the high wind sequence. According to experience, the threshold size must be established such that there are specific years within the data period devoid of typhoons, and the years without typhoons should not surpass one-tenth of the total years. Table 2.2 illustrates the occurrence of severe wind speeds at an elevation of 70 m along the southeast coast.

2. **Turbulence**

Turbulence Intensity (TI) denotes the extent of stochastic fluctuations in wind velocity during a 10-min period. It is defined as the ratio of the standard deviation of the 10-min average wind speed to the mean wind speed during the same interval. This characteristic delineates the wind conditions of a wind farm in a specific location

Table 2.2 The frequency of extreme wind speeds at a height of 70 m along the southeast coast

Wind speed (m/s)	Number of stations	Frequency (%)	Cumulative frequency (%)
< 37.5	48	22.75	22.75
37.5–42.4	42	19.91	42.65
42.5–49.9	89	42.18	84.83
50–51.9	16	7.58	92.42
52–53.9	9	4.27	96.68
54–55.9	5	2.37	99.05
56–57.9	1	0.47	99.53
58–59.9	1	0.47	100

and serves as a critical indication for the safety classification and design criteria of wind turbines as outlined in IEC 61400-1.

The primary causes of turbulence are twofold: firstly, air flow is influenced by friction or drag from surface irregularities; secondly, vertical air motions are generated by variations in air density and ambient temperature. Turbulence is often induced by a combination of these two primary sources. Under neutral atmospheric circumstances, air experiences adiabatic cooling as it ascends, achieving thermal equilibrium with the surrounding environment. Consequently, in a neutral environment, the turbulence intensity predominantly relies on the surface roughness.

Turbulence can diminish the efficiency of wind energy consumption in turbines, lower power production, and elevate the fatigue loads and wear probabilities of turbine components. Wind turbulence can induce random forced vibrations in wind turbines with flexible frames. In linear structural systems, an increase in turbulence intensity by 2 to 3 times might lead to a comparable rise in the structural dynamic response or fluctuating wind loads. The distinct turbulence characteristics of the near-surface layer at the core of a typhoon significantly affect the safe operation of wind turbines compared to onshore circumstances. Intense turbulence during a typhoon frequently serves as a principal factor in vibration-induced breakdowns of wind turbines.

The standard turbulence design parameters for wind turbines often do not surpass 0.2, which is appropriate for wind farms located in regions devoid of typhoon influences and characterized by smooth terrain. Nonetheless, in offshore wind farms, typhoons represent a critical aspect that must not be overlooked. The wind turbine's aerodynamic resistance design requires further study and testing to accommodate harsh situations. The turbulence intensity at the core of a typhoon can attain values between 0.6 and 0.9, in contrast to the typical turbulence intensity of around 0.1 under non-typhoon circumstances, leading to significant stresses on wind turbines [8].

3. **Sudden wind direction change**

During a typhoon's passage through a wind farm, the wind direction may alter swiftly within a few timeframe. This poses a significant challenge to the wind turbine yaw system's capacity for rapid response to mitigate overall loads, as well as the turbine's safety during the typhoon. Consequently, it is vital to comprehend the swift alterations in wind direction throughout several directional ranges of the typhoon.

As the typhoon's core traverses, the pressure curve manifests a funnel form, characterized by a rapid increase and decrease in air pressure. The wind direction at all measurement places can fluctuate by over 45° in a brief interval, with locations nearer to the center undergoing abrupt changes of 120° to 180°, transitioning from northeast winds to south or southwest winds.

Upon the landfall of a typhoon, the external power grid of the wind farm may sustain damage, inhibiting the wind turbine's ability to adapt to the wind, hence precluding its operation in accordance with the established control plan. The wind direction fluctuates incessantly with the progression of the typhoon's core. In pitch-controlled wind turbines that are already inoperative, an abrupt alteration in wind direction results in the principal wind vector transitioning from directly front of the turbine to the lateral and posterior positions. As a result, the effective wind area of the entire wind turbine fluctuates, causing persistent high winds to impact the blades from many directions. This produces sustained severe bending and torsional moments. The wind speed during a typhoon may surpass design thresholds, potentially causing blade fractures and damage to the pitch mechanism. Under standard conditions, lateral winds on a wind turbine result in maximal stresses, potentially causing the fracture of tower connection bolts and yaw deformation. Consequently, due to the interplay of static and dynamic forces, fatigue loads progressively build, ultimately approaching or above the design load thresholds of the tower. In extreme instances, this may lead to the tower's collapse [9].

4. **Typhoon waves**

Within the typhoon's vicinity, the sea surface will produce colossal waves. During the early phases of the typhoon, despite elevated wind velocities, wave heights remain relatively modest, often about 3 m in the peripheral areas, although toward the center, they may attain heights of 4–5 m. As the typhoon progresses, wave heights escalate alongside increasing wind speeds, demonstrating a substantial positive link between the two variables. Upon reaching its mature stage, the typhoon's wind speed stabilizes, but the region of intense winds progressively extends outward. Consequently, the wave heights, having attained their peak, cease to rise, and the expanse of substantial waves progressively expands into the peripheral areas of the typhoon. As the typhoon starts to fade or weaken, despite a reduction in wind speed, the impacted maritime regions frequently continue to see considerable wave heights.

After the substantial waves in the typhoon's center have developed, they radiate outward from the epicenter. As these substantial waves exit the typhoon region and go outward, they transform into swell waves. Swell waves radiate outward at a velocity of 2 to 3 times that of the typhoon. The orientation of the swell signifies the location of the typhoon's core. If the swell direction remains unchanged as the wave height progressively rises, it signifies the imminent arrival of a typhoon.

Offshore wind farms are often impacted by typhoons for a duration of 12 to 24 h, and the prolonged interaction of wind and waves during a typhoon can result in low-cycle fatigue damage to wind turbines. In contrast to typical sea waves, the waves in a typhoon zone are considerably bigger and exert much greater impact forces, resulting in more severe impacts on wind turbines [10]. Moreover, inside the typhoon vortex region, intense precipitation frequently coincides with elevated wave activity, which may impact the blades and may result in blade damage.

2.2 Marine Hydrological Environment

The offshore natural environment is severe, characterized by the interplay of wind, waves, and currents, which substantially complicates the design of wind turbines and the construction of offshore wind farms. The maritime hydrological environment is essential for the design and installation of wind turbines, and comprehending this environment is a necessity for offshore wind development. This section will succinctly present the marine hydrological environment from three perspectives: ocean currents, waves, and sea ice.

2.2.1 *Ocean Currents*

Ocean currents denote the displacement of water masses from one site to another within the marine environment. They often denote extensive, rather constant, non-periodic water flows in both horizontal and vertical orientations, constituting one of the primary kinds of seawater movement. The phrase "large-scale" denotes fluxes that transpire over extensive spatial dimensions, including hundreds to thousands of kilometers, or perhaps a global extent. The term "relatively stable" signifies that, over extended durations—such as a month, season, year, or many years—the trajectory, velocity, and course of the flow stay largely consistent.

The primary causes of ocean current formation are twofold: first, wind-driven currents, which arise from the influence of surface winds; second, density currents, also termed thermohaline circulation, which result from uneven heating and cooling of the ocean surface, as well as disparities in evaporation and precipitation, leading to variations in temperature and salinity due to density differences. The forces influencing seawater can be categorized into two primary types: the first encompasses forces that induce seawater movement, including gravity, pressure gradient forces, wind stress, and tidal forces; the second comprises forces resulting from seawater movement, such as Coriolis force, shear stress, and frictional forces.

Upon the completion of offshore wind turbine foundation construction, the disruption of surface soil and the establishment of permanent impediments will markedly influence the flow of water particles induced by currents and waves. A horseshoe-shaped vortex will first develop in front of the foundation. Secondly, eddies will develop on the rear side of the wind turbine tower. The flow lines adjacent to the wind turbine foundation will converge. The alteration in local flow dynamics will elevate the shear stress of the water flow on the bottom, resulting in an increased sediment transport capacity. If the seafloor is prone to erosion, localized scouring pits may develop around the foundation. Scouring pits might compromise the stability of the foundation. Consequently, it is imperative to monitor and evaluate the currents when engineering offshore wind turbines [11].

The existing measurement stations have to be categorized into long-term current measurement stations and comprehensive tidal hydrology measurement stations. The observation methodologies and durations must adhere to the "Specifications for Wind Energy Resource Measurement and Ocean Hydrological Observation of Offshore Wind Farms" NB/T 31029-2019. For comprehensive tidal hydrological measurement stations, the designated observation times must be scheduled in accordance with the predicted tidal timings of the tidal stations inside the engineering maritime zone and modified as necessary based on the actual tide circumstances. Observations are inadvisable when wind force surpasses scale 6 [12].

The requirements for current observation are as follows:

- The configuration of the full-tide hydrological measurement stations must include the engineering area and be constructed in diverse channels, bays, and estuaries that significantly influence the tidal current characteristics of the engineering zone.
- In wind farms characterized by intricate seabed topography, coasts, and dynamic settings, the quantity of full-tide hydrological monitoring stations should be suitably augmented in accordance with the site's extent.

The components of the current observation must encompass flow velocity and orientation, augmented by assessments of water depth, transient wind speed, and wind direction. The resolution for flow speed observation must not surpass 1 cm/s, and the resolution for flow direction observation must not exceed 1°. The precision of flow velocity and directional measurements must adhere to the criteria specified in Table 2.3.

The layers for the present observation must be stratified based on the recorded water depth and adhere to the specifications stated in Table 2.4.

Continuous monitoring of currents must occur at a minimum frequency of once per hour. During intervals of swift increases, decreases, or fluctuations in flow, supplementary measurements should be conducted every thirty minutes. The recorded values for flow speed and direction should preferably represent the average of 100 s of measurements; if not, the duration of the sampling interval must be specified in the observation records.

Table 2.3 Accuracy of flow speed and direction observation

Flow speed (cm/s)	Flow speed accuracy	Flow direction accuracy
$v < 100$	± 5 cm/s	$\pm 5°$
$v \geq 100$	$\pm 5\%v$	

Table 2.4 Total water depth and observation layers

Total depth	Observation layers
$D \leq 5$ m	0.2, 0.6, and 0.8 D
5 m < $D \leq 50$ m	Surface layer, 0.2, 0.4, 0.6, and 0.8 D, bottom layer
50 m < $D \leq 100$ m	Surface layer, 5, 10, 15, 25, 30, 50, and 75 m, bottom layer

Note The surface layer denotes the water layer situated 0.5 m beneath the ocean surface. The bottom layer denotes the water stratum situated 0.5 m above the seafloor. 0.2, 0.4, 0.6, and 0.8 D denote the water layers situated 0.2, 0.4, 0.6, and 0.8 D beneath the water surface, with D being the total water depth. The depths of 5, 10, 15, 25, 30, 50, and 75 m denote the respective measurements below the ocean surface. When examining the bottom layer, it is crucial to ensure that the device does not contact the seafloor

Current observations should ideally employ methods such as anchored ship measurements, anchored subsurface buoy measurements, and anchored surface buoy measurements for continuous point assessments. The observational devices must comprise direct-reading current meters and acoustic Doppler current profilers.

2.2.2 *Waves*

Ocean waves represent a significant mode of saltwater flow, extending from the surface to the depths of the ocean. Their development results from the interplay of external forces, gravitational influence, and the surface tension of water. Numerous variables can induce waves on the ocean surface, including wind, variations in air pressure, tidal influences from celestial bodies, and undersea seismic activity. The essential attribute of waves is that, when subjected to diverse external forces, water particles oscillate regularly from their equilibrium locations, leading to the transmission of wave shapes.

Wave observation should meet the following requirements.

(1) The parameters for wave observation in offshore wind farms must encompass wave height, wave period, and wave direction. The resolution for wave height measurement must not surpass 0.1 m, for wave period measurement must not exceed 0.1 s, and for wave direction measurement must not exceed 1°. The precision of wave observations must adhere to the standards indicated in Table 2.5.

Table 2.5 Accuracy of wave observation

Project		Accuracy
Wave height	$H \leq 1$m	± 0.1m
	H>1m	± 10%H
Wave period		± 0.5 s
Wave direction		± 5°

(2) Continuous wave observations must be performed at a minimum frequency of once per hour. The sample time interval must not surpass 0.5 s, and the quantity of continuously recorded waves should be no fewer than 100. The recording time must be established according to the average wave period and may be configured between 17 and 20 min.

Wave observations should ideally employ devices for detecting gravity, pressure, acoustic, or ultrasonic waves. The selection of the suitable wave measurement instrument must be based on project specifications and the sea conditions at the observation location, with the technique of securing the device established correspondingly.

2.2.3 Sea Ice

All ice present in marine environments is generally termed sea ice, encompassing not only ice created by the direct congelation of saltwater but also ice from rivers, lakes, and glaciers derived from terrestrial sources. The consequences of sea ice, including obstruction of trade routes, damage and sinking of vessels, and impairment of infrastructure, are commonly referred to as sea ice catastrophes. Sea ice catastrophes represent a significant maritime danger in polar regions and certain high-latitude locales. The detrimental impact of sea ice in marine ecosystems is considerable, particularly with drift ice and icebergs. Large expanses of drift ice can shift collectively due to ocean currents and wind, exerting pressure on the bases of wind turbines. Free-floating drift ice can impact the foundations, inducing vibrations that compromise the safety of wind turbines. Furthermore, sea ice presents a considerable risk to the safety of vessels used in the installation and maintenance of offshore wind turbines. Consequently, it is important to evaluate the influence of sea ice in the design of offshore wind turbines.

During "light ice years" or "moderately light ice years," sea ice typically exerts minimal influence on offshore operations, or it only marginally inhibits the utilization of northern ice-locked ports. Nevertheless, if the circumstances for ice formation attain "perennial ice years" or more severe classifications, particularly during "extreme ice years," catastrophic events may ensue.

Sea ice disasters result from extended durations of low temperatures persisting over several days, weeks, or even throughout winter. The duration of a catastrophic occurrence can vary considerably, ranging from a few days to several weeks, and in certain instances, over two months.

The primary observation elements and supplementary observation components for sea ice should encompass the following:

- The primary visible properties of floating ice are volume, density, type, surface features, condition, drift direction and velocity, thickness, and the perimeter of the ice region.
- The primary visible characteristics of permanent ice consist of ice type, ice thickness, and ice border.

Table 2.6 Accuracy of sea ice observation elements

Observation elements	Accuracy
Ice quantity, density	± 10%
Drift direction	± 5°
Drift speed	± 0.1 m/s
Ice thickness	± 1 cm

- Additional observational criteria for sea ice encompass sea surface visibility, air temperature, wind velocity, wind orientation, and meteorological events.

The precision of the sea ice observation components must adhere to the criteria specified in Table 2.6.

Sea ice observations must be conducted from a survey vessel, with data collection commencing upon the vessel's arrival at the station. The observation point for sea ice must be designated at an elevated location on the vessel. The major focus of observation should be on sea ice located beyond twice the length of the vessel to minimize the survey vessel's influence on ice assessments.

2.3 Marine Geology

The marine geological attributes significantly affect the design of offshore wind turbine foundations. This section will concisely analyze various marine geological conditions.

2.3.1 Marine Sediments

Marine sediments comprise inorganic materials, like mud and sand, as well as organic matter, such as biological residues. These materials settle on the seabed as a result of the interplay between gravitational forces and marine dynamics, including water currents. Rocks in the Earth's crust undergo weathering and erosion by physical, chemical, and biological processes, leading to substantial debris buildup. These particles are transported into the ocean by rivers, precipitation, glaciers, and wind. Certain debris accumulates on the continental shelf, while others are carried by ocean currents to the abyssal zone. Sediments derived from terrestrial sources are referred to as terrigenous sediments. The ocean teems with life, and the remains of perished marine organisms collect on the seabed, forming deep-sea sediments mostly composed of siliceous and calcareous oozes. These are termed biogenic sediments. Volcanic eruptions release significant amounts of magma, ash, and volcanic debris into water, producing volcanic sediments. Cosmic events, such as planetary motion and collisions, generate cosmic dust, with around 14 tons of this material falling to Earth daily, most of which settles on the ocean floor. The resulting deposits are referred to as cosmic sediments. Chemical reactions in seawater produce chemical precipitates referred to as authigenic sediments [13].

Sediments have components such as rocks, gravel, sand, and clay, and are often classified and described based on their median particle size. Under the Wentworth classification system, sediments with particle sizes over 256 mm are classified as rocks. Particles ranging from 2 to 256 mm are classified as gravel, and those from 0.5 to 2.0 mm are identified as coarse sand. Medium sand refers to sediments with particle sizes between 0.25 and 0.5 mm, whilst fine sand applies to those with particle sizes ranging from 0.065 to 0.25 mm. Coarse silt includes sediments with particle sizes from 0.016 to 0.065 mm, fine silt refers to those between 0.004 and 0.016 mm, while sediments smaller than 0.004 mm are categorized as clay. Geological conditions vary throughout different marine regions. In China's eastern coastal regions, the substrate mostly consists of significant deposits of silt, generally composed of fine sand, silt, and clay. The selection of wind turbine foundation types must be based on the geological attributes of the seabed and the depth of the water [14].

2.3.2 Classification of Marine Areas

Marine areas in China can be classified, from nearshore to offshore, into tidal flats, the intertidal zone, nearshore waters, and offshore seas. Each of these marine zones possesses unique geological characteristics and differing water depths.

1. **Tidal Flats**

Tidal flats is the conventional word used in China to refer to mud-rich sedimentary coastal areas, lake edges, and riverbanks. Coastal tidal flats are often divided into three parts. Supratidal flats refer to the sedimentary area located above the mean high water line of the typical spring tide, predominantly composed of silty and clayey sediments. The second refers to the intertidal flats, which are the areas situated between the mean high water level and the mean low water line during the average spring tide. The substrate in this region predominantly comprises muddy, sandy, and rocky components. The third is the subtidal flats, indicating the sedimentary area below the mean low water mark. The ground in this area primarily comprises silty sand.

2. **Intertidal Zone**

The intertidal zone is the area between the high tide line and the low tide line, sometimes referred to as "marine mudflat." The water depth in this region frequently does not surpass 5 m [15]. The intertidal zone is divided into three regions based on tidal activity. The high tide zone is the top zone. Located at the uppermost region of the intertidal zone, with the upper boundary defined by the mean high water line of the spring tide and the lower boundary indicated by the mean high water line of the neap tide. This area is temporarily flooded, often only during spring tides. The second zone is the mid-tide zone. This region includes the majority of the intertidal zone, with its upper boundary marked by the mean high water line of the neap tide and its lower boundary defined by the mean low water line of the neap tide. It is a distinctive intertidal zone. The third zone is the infralittoral zone. This is the lowest stratum of the intertidal zone, delineated by the upper boundary at the mean low water line of the neap tide and the lower boundary at the mean low water line of the spring tide. This area is primarily submerged, becoming accessible to the atmosphere only during the short period of ebb tide linked to spring tides.

3. **Nearshore and Offshore**

Nearshore refers to the maritime area with a water depth between 5 and 50 m below the lowest tide level. Offshore refers to the nautical area when the sea depth surpasses 50 m. The seabed composition in these areas mostly comprises rock, silt, and clay [16].

2.4 Other Marine Environments

2.4.1 Climatic Environment

1. **Seawater Temperature**

The temperature of seawater is a crucial physical and mechanical characteristic of saltwater. It directly affects several physical and dynamic properties of saltwater and acts as a principal indication of the average kinetic energy of molecular thermal motion in seawater. Furthermore, it functions as an essential parameter for evaluating the thermal energy of saltwater.

The Bohai Sea and northern Yellow Sea in China are markedly affected by the continental climate, resulting in large seasonal variations in seawater temperature. The temperature of seawater in the southern Yellow Sea and the East China Sea is closely associated with the distribution of ocean currents and water masses. The seawater temperature in the South China Sea exhibits characteristics of a tropical deep sea, marked by continuously elevated temperatures year-round, with little regional variations and seasonal changes.

The distribution of seawater temperature in China's coastal waters may be categorized into three types: winter type, summer type, and transitional type. The winter type spans from November to March of the following year, indicating the yearly nadir of ocean temperatures. During this period, surface seawater temperatures surpass air temperatures, but land temperatures are lower than sea temperatures. As a result, coastal seawater temperatures are reduced, whilst offshore seawater temperatures are increased. In the winter type, the horizontal temperature gradient is significant, isotherms are closely aligned, and their configuration is frequently parallel to the coastline, with surface seawater temperatures increasing from north to south. The summer season occurs from June to August, marked by heightened solar radiation that leads to increased surface seawater temperatures in China's coastal waters, making it the warmest season for seawater temperatures each year. During this period, coastal seawater temperatures surpass those of offshore waters due to increased air temperatures, resulting in a more uniform distribution of seawater temperatures, a diminished horizontal temperature gradient, and irregular isotherm distribution, with minimal temperature variations between the north and south. The transitional type occurs throughout the seasonal changes of April to May and September to October, with spring representing a warming phase and autumn denoting a cooling phase. The defining characteristic of the transitional type is its complex and variable temperature conditions, which are unstable and less predictable.

The vertical distribution of seawater temperature in China's coastal waters is significantly influenced by meteorological conditions. In winter, it is mostly affected by variable polar continental air masses, with the sea surface frequently subjected to strong northerly winds. This results in heat loss from the ocean surface, lowering the surface seawater temperature, increasing density, and promoting convective mixing between the top and bottom water layers. The vertical distribution of seawater temperature generally equilibrates within the depth reached by this mixing. As winter temperatures decline, heat loss from the sea surface escalates, enhancing vertical convection and increasing the mixing area. Consequently, throughout winter, the seawater temperature in shallow areas frequently remains uniform from the surface to the seabed. The specific date of the formation of the uniform seawater temperature layer in shallow areas varies across different coastal regions in China. In the Bohai Sea, it is seen from October to March of the following year; in the Yellow Sea, from November to April; in the shallow areas of the East China Sea continental shelf, from December to April; and in the shallow parts of the northern South China Sea, from December to March. Furthermore, stable seawater temperature layers may form in the deep areas of the East China Sea and South China Sea at depths between 75 and 150 m. The formation and duration of uniform seawater temperature strata vary by maritime area, with northern waters demonstrating early development and extended persistence, whereas southern seas demonstrate later formation and shorter durations.

Seawater temperature measurements must be conducted simultaneously with current observations during the whole tidal hydrological survey period. The establishment of measurement stations must comply with the following regulations: The observation duration and intervals must align with the current observations during the extensive tidal hydrological survey; the resolution of seawater temperature readings should not exceed 0.05 °C, and the accuracy must be ± 0.2 °C. Continuous point measurements of seawater temperature should be conducted using a Conductivity-Temperature-Depth (CTD) instrument. CTDs are categorized into two main types: real-time display and self-recording.

2. **Humidity**

Atmospheric humidity, sometimes referred to as humidity, is a physical measurement indicating the amount of water vapor present in the atmosphere. Humidity may be quantified in several manners, with commonly employed marine measurements being vapor pressure, absolute humidity, relative humidity, and saturation deficit.

Atmospheric humidity affects meteorological events, such as cloud formation, precipitation, and salt fog density, and it also determines the air saturation level within a cabin. The humidity in a maritime environment is affected by several factors, including temperature, duration of sunlight, wind speed, and wave conditions. The relative humidity above the ocean is markedly high, and the atmosphere is characterized by persistently higher relative humidity levels.

Humidity often accompanies various climatic conditions and generally acts as a catalyst for issues such as mold growth, enhanced corrosion, and intensified electrochemical processes. In the South China Sea region of our country, increased relative humidity often coincides with increasing temperatures. The southeastern coastal regions of China have hot and humid summers, marked by a prolonged rainy season and an average monthly relative humidity of 85%. In this environment, water vapor can infiltrate the interiors of enclosures with an IP54 protection grade and enclosed insulating components, leading to performance complications [17].

2.4.2 *Salt Mist*

Salt mist is a dispersed system of minuscule liquid droplets containing sodium chloride (NaCl) suspended in the atmosphere, constituting a crucial element of marine atmospheric dynamics. Coastal areas and offshore air are infused with salt due to the evaporation of seawater, which dissolves in tiny water droplets, creating a highly concentrated salt mist. The primary agent of corrosion damage resulting from salt mist is the chloride ion (Cl^-). The Cl^- ion has a small radius of 1.81×10^{-10} m, granting it considerable penetrating powers, allowing it to traverse the metal oxide layer and protective coatings to reach the metal core, so undermining the metal's passive state [18].

2.4.3 *Lightning*

Coastal regions in China sometimes encounter intense convective weather, marked by increased humidity and enhanced salinity, leading to a more severe atmospheric condition than that of terrestrial places. This scenario markedly elevates the probability of ground flashes, potentially resulting in considerable damage to onshore wind turbine systems. Data indicate that losses from lightning strikes on coastal wind turbines are over 80% of all lightning strike occurrences in the sector [19]. Given the inherently unpredictable nature of lightning phenomena, it is impracticable to entirely prevent lightning strikes on wind turbines. Consequently, it is imperative to examine the damage mechanisms of lightning strikes and incorporate lightning protection technologies into the design, production, and installation stages of offshore wind turbines. This will mitigate the losses incurred by lightning strikes [20].

References

1. Policy and Regulation Department of the China Meteorological Administration. Tropical Cyclone Classification: GB/T19201—2006. Beijing: China Standards Press; 2006.
2. Zhang C. Analysis of the risk characteristics of typhoon disasters landing on China's coast. Ganzhou: Jiangxi University of Science and Technology.
3. Su H, Dong G, Wang M, et al. Analysis of the main characteristics and causes of disasters of typhoons landing from 1949 to 2018. Environ Sci Manag. 2020;45(5):4.
4. Xie B, Zeng C. Statistical characteristics of tropical cyclones landing in Guangdong. Guangdong Meteorol. 1998;3:3.
5. Hu Y, Song L, Liu A. Comparison of the climatic characteristics of tropical cyclones landing in different regions of China. Atmos Sci Res Appl. 2008;000(001):1–8.
6. Song L, Mao H, Qian G, et al. Analysis of the impact of tropical cyclones on wind power generation. J Solar Energy 2006;09:961–5.
7. Chen Q. Key technology research on typhoon-type wind turbines. Hangzhou: Zhejiang University; 2017.
8. Wu J. Design of offshore wind turbines. Beijing: Chemical Industry Press; 2012.
9. Wang H. Typhoon disaster protection for wind farms. Beijing: Chemical Industry Press; 2017.
10. Wang W. Foundation design of offshore meteorological towers. Beijing: China Water Power Press; 2016.
11. China Electricity Council. GB/T36569—2018. State administration for market regulation; Standardization Administration of China; 2018.
12. Wind Power Standardization Technical Committee of the Energy Industry, Wind Farm Planning and Design Group. Specification for wind energy resource measurement and oceanographic observation of offshore wind farms: NB/T31029—2019. Beijing: China Water Power Press; 2020.
13. Zhao S. Marine environmental science. Beijing: Ocean Press; 2011.
14. Yang Z. Marine geology. Shandong: Shandong Education Press; 2004.
15. Zeng Y. Marine engineering environment. Shanghai: Shanghai Jiao Tong University Press; 2007.
16. Chen X. Design and development of offshore wind turbines. China Electric Power Press; 2018.
17. National Anti-Corrosion Standardization Technical Committee. Anti-corrosion technical specifications for coastal and offshore wind turbine units: GB/T33423—2016. Beijing: China Standards Press; 2016.
18. Ma A. Corrosion prevention engineering of offshore wind farms. Beijing: China Water Power Press; 2015.
19. Wang R, Yang H, Li B, et al. Characteristics of lightning strikes and lightning protection measures in offshore wind farms. In: Proceedings of the 4th China wind power post-market symposium, 2017.
20. Zhao H, Wang X. Lightning strike mechanism and lightning protection technology of wind turbines. Power Syst Technol. 2003;27(7):5.

Chapter 3
Offshore Wind Turbine Design Technology

The wind turbine design is the basis of reliable and economical operation. The design method of offshore wind turbines is different from that of onshore wind turbines because of their more severe condition in complex environments. This chapter will introduce the integrated design, reliability design, typhoon-proof design, corrosion-proof design and lightning-proof design of offshore wind turbines.

3.1 Integrated Design

3.1.1 Content and Significance

Integration can be understood as a measure to integrate two or more different and uncoordinated matters into a whole by adopting appropriate ways, methods or measures to form a synergistic effect to achieve the goal of organizational planning [1]. The concept of integration has been widely applied in various fields, such as economic integration, political integration, urban–rural integration, producing-transport-marketing integration, electro-mechanical integrating, integrated design, etc. Through module integration, the cost can be effectively reduced and the efficiency can be improved. The design of offshore wind turbines involves many complicated modules, such as resource evaluation, foundation design, wind turbine design, electrical design, monitoring and control system design, etc. High cost is the key to restricting the development of offshore wind power. Introducing the concept of integration into the design of offshore wind turbines is expected to greatly reduce the development cost and promote the healthy and rapid development of offshore wind power.

At present, the integrated design of offshore wind turbines mainly includes the integrated design of wind turbine-tower-foundation, blade pneumatic-structure-load, electrical system, offshore wind farm integrated monitoring system, offshore

M. Ge et al., *Offshore Wind Turbine Technology*,
https://doi.org/10.1007/978-981-96-2175-0_3

wind farm natural conditions and wind turbine type selection, offshore wind power construction integration, offshore wind farm operation integration, etc.

In the design approach of offshore wind turbines, the wind turbine-tower-foundation integrated design is the most important and has received the most extensive research. It has already been practically applied in numerous offshore wind turbine designs both domestically and internationally. Therefore, this section will primarily explain the wind turbine-tower-foundation integrated design method. Any mention of integrated design hereafter refers specifically to the wind turbine-tower-foundation integrated design.

The integrated design of offshore wind turbines is to take the external environmental conditions such as engine room, tower, foundation and wind condition, sea condition, and seabed geology of offshore wind turbines as a unified whole dynamical system for load simulation analysis, structural verification and optimization. It can comprehensively evaluate the force status of the offshore wind turbine, improve the design safety, reduce the repeated calculation work, improve the design efficiency, effectively optimize the structure, and reduce the quality and cost of the supporting structure [2].

3.1.2 Traditional Design and Integrated Design Contrast

1. **Traditional Design**

The traditional design of offshore wind turbines is a step-by-step iterative design with the bottom of the tower as the interface, as shown in Fig. 3.1. Based on the basic data, the wind turbine-tower load is calculated by the wind turbine manufacturers, and the limit load at the bottom flange of the tower is extracted. According to the ultimate load of the flange at the bottom of the tower, the corresponding foundation is designed by the qualified design unit. The two parts are designed and optimized respectively, then the offshore wind turbine and its foundation design are finally completed [3]. In traditional design, the tower and foundation are designed separately, each considering its factor of safety. The maximum wind load and the limit wave load are superimposed, which makes the design relatively conservative. The wind-induced fatigue and wave-induced fatigue are added, which makes the results inaccurate. The wave load is calculated repeatedly to a certain extent, which makes the result larger. To sum up, although the traditional design has a certain effect, but relatively conservative, low efficiency, and high cost. From the perspective of integrated design to find the optimal solution for the overall structure, there is still much room for optimization and expansion.

2. **Integrated Design**

Integrated offshore wind turbine design is a two-step integration of traditional design, as shown in Fig. 3.2. This method takes the wind turbine and the foundation as a whole, considering the different environmental and external conditions of the

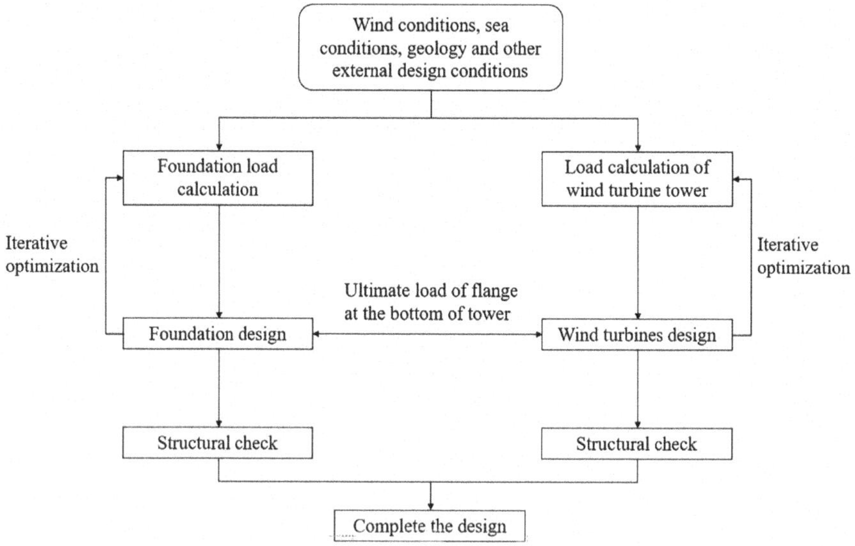

Fig. 3.1 Step by step iterative design

offshore wind farm. And the basic structure is designed and verified. It mainly includes integrated modeling, integrated load analysis and integrated structural verification.

Integrated design usually involves the following steps:

- Foundation selection and preliminary design;
- Load estimation;
- Load modeling, including wind conditions modeling, sea conditions modeling and supporting structure modeling;
- Load simulation is carried out by combining wind and sea conditions;
- The fatigue load and extreme load of each interface are extracted and processed;
- Structure check and optimization design.

3. **The Difference Between Integrated Design and Step-by-Step Iterative Design**

(1) Load mode

The coupling effect of wind and wave is considered in the load analysis of integrated design, and the load obtained can be directly applied to structural design verification. Based on calculating the load, the step-by-step iterative design should superimpose the wave load according to the relevant design specification.

(2) Safety standards

The integrated design unifies the tower and foundation design. It takes the same safety factor and safety criterion according to the international standard IEC offshore wind

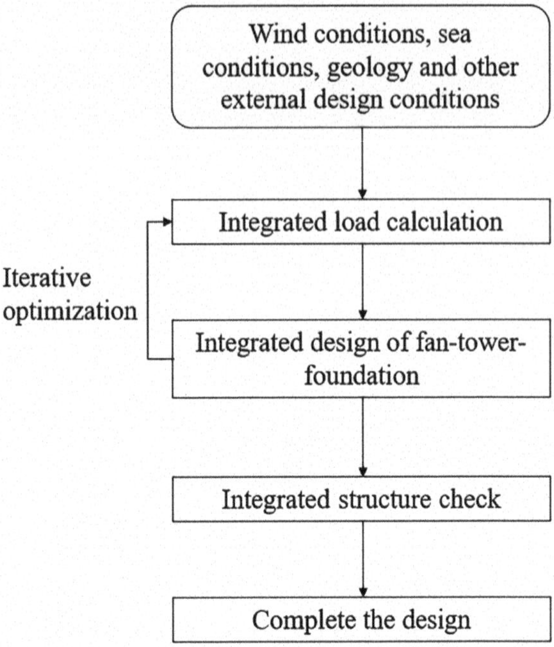

Fig. 3.2 Integrated design

turbine design code. The step-by-step iterative design separates the tower and the foundation. It considers the safety factor separately. The overall safety factor is larger.

(3) Fatigue analysis

In the integrated design, the coupling load obtained from the integrated load analysis is used for the structural fatigue analysis. The step-by-step iterative design usually superposes the fatigue damage of wind load and wave load linearly. The result is not accurate [4].

3.1.3 Advantages and Difficulties

1. **Advantages**

Compared with traditional design, integrated design has the following advantages:

- The interactive information of the traditional design should be transmitted through the interface load of the tower's bottom flange. In the process of transmission, the working condition information may be lost, which may cause unreasonable superposition of extreme working conditions and make the design too conservative. The integrated design uses the whole design. It does not need the information interaction and does not exist this kind of question.

- In the traditional design process, the wind turbine manufacturer needs to consider the wind condition and sea condition when calculating the interface load. The foundation design entity cannot decompose the interface load. The interface load and the wave load can only act on the foundation at the same time, resulting in a repeated calculation of wave load to a certain extent, which makes the result larger. The integrated design directly regards the wind turbine, tower, foundation and external environmental conditions as an entirety, which can avoid this problem.
- The traditional design is only based on the interface load. If extreme force is applied to the system in the form of a static load, the result will be very different from the real state. The integrated design uses integrated load simulation, which can accurately simulate the dynamic changes of the load and make the results more accurate [2].
- The traditional design separates the tower and foundation, and the load simulation calculation may use different simulation software. According to different design standards, there is a possibility of conflict between software and software, and between standards and standards, which increases the calculation error. The integrated design adopts unified software and design standards, and the calculation error is relatively small.

Reference [2] gives the simulation results of a 5.5 MW wind turbine with traditional design and integrated design, as shown in Table 3.1. The table shows that the integrated design has a significant optimization effect.

Thus, an integrated design can:

- Reducing Costs. According to statistics, the material and installation costs of the offshore wind turbine support structure account for nearly 30% of the total cost. Reducing the tower and foundation costs can significantly lower the levelized cost of energy of offshore wind turbines. Integrated design achieves overall optimization, effectively reducing the material requirements for towers and foundations, enabling optimized design and manufacturing of offshore wind turbine support structures. This leads to lower manufacturing costs and enhances the competitiveness of domestic offshore wind turbines.

Table 3.1 The simulation results of traditional design and integrated design

Part	Parameter	The degree of decline (%)
Tower bottom load	Ultimate load	19.7
	Fatigue load	9.8
Tower	Wall thickness (average)	10.5
	Overall quality	10.4
Bearing platform	Outer diameter	3.3
	Quality	6.6
Steel pipe pile	Wall thickness	2.5
	Quality	2.5

- Improve security. The integrated design can comprehensively evaluate the overall stress state of offshore wind turbines. It can select the same safety factor and safety standard for wind turbines and their foundations, which can effectively improve the design safety.
- Promote collaborative development. The integrated design breaks the previous design pattern. It makes the wind turbine manufacturers and design units better cooperate to improve design efficiency. It can reduce the amount of work and promote the healthy development of the offshore wind power industry.

2. **Difficulties**

(1) Lack of uniform standards

At present, there are many design standards for offshore wind turbines. There are international standard IEC61400 series for offshore wind turbine design code, load code for port engineering JTJ215-98, code for pile foundation for harbor engineering JTS167-4 and code for design of wind turbine foundations for offshore wind power projects NB/T10105, etc. From the view of the overall design, the IEC International Standard puts clear requirements for the tower and foundation design of offshore wind turbines. The domestic harbor standard and industry standard mainly aim at the traditional design. There are some conflicts with the international standard in the requirements and indexes. Speeding up the unification of standards and carrying out the standards suitable for integrated design are the basis for the practical application of integrated design of offshore wind turbines [2].

(2) Integrated modeling and simulation are difficult

Offshore wind turbines, foundations, wind conditions, sea conditions, seabed geology and other external environmental conditions are a unified whole. The integrated modeling and simulation of this whole is the most basic requirement of integrated design. The focus of the integrated design is not whether the entire modeling and simulation are carried out, but whether the overall dynamic response of the wind turbine and its foundation is fully considered according to the overall modeling and simulation results. Whether the overall adjustment and global optimization of the design optimization are carried out is also considered. Only wind turbine manufacturers continue to improve their research and development capabilities, open cooperation between design units, and third parties play a role in intellectual property protection and coordination, global optimization of offshore wind turbines is more likely to be achieved [2].

(3) Design risk and return are not equal

At present, the design risks and benefits of offshore wind power supply chains are not equal. See Table 3.2. The DNV GL classification society has made statistics on the benefits of the optimization of wind turbines to reduce costs. Although wind turbine designers have to bear more design costs and risks to reduce costs, the foundation designers can get more cost-cutting benefits than the wind turbine designers. Only through the joint efforts of the market, the government and the enterprises to solve

Table 3.2 Optimize cost–benefit reduction statistics everywhere

Optimization methods	The degree of cost reduction	
	Wind turbine (%)	Foundation (%)
Integrated design	0.2	1.8
Optimize blades	1.1	1.1
Enhanced control strategy	1.4	1.1
Relax frequency constraints	0.5	4.7

this contradiction, can the wind turbine manufacturers and the foundation designer entities cooperate better. It will be conducive to better implementing the integrated design.

3.1.4 Application

Since 1991, when the world's first offshore wind farm Vindeby was put into operation in Denmark, Europe has continuously deepened its research on the design of offshore wind turbine systems. Integrated design initially originated from the optimization of offshore wind turbine systems in Europe. Internationally, research on integrated design, particularly in load simulation, has yielded abundant results [5, 6]. Many companies and organizations are working to apply integrated design thinking to the design process. The International Energy Agency (IEA) launched a series of research projects, OC3, OC4, and OC5 (Offshore Code Comparison Collaboration Continued with Correlation) [7], aiming to improve the accuracy of offshore wind turbine load simulation tools and to develop an integrated design analysis platform for offshore wind turbines. DNV GL classification society put forward "Project FORCE (For Reduced Cost of Energy)", which includes integration design, enhanced control strategy, optimized blade and relaxed frequency constraints. Among them, integrated design refers to the numerical modeling of the wind turbine-tower-foundation as a whole for comprehensive coupling. It can not only make the support structure have higher stiffness, but also reduce the cost. Through the joint operation of various optimization methods, it can reduce the levelized cost of energy (LCOE) for offshore wind turbines by at least 10%. At the same time, DNV GL has also analyzed the obstacles to the implementation of integrated design. It proposed determined by market forces, buyer-led enforcement, government-led information sharing and implementation of design, engineering and procurement integration of JIP (Joint Industry Projects) solutions, etc. And the timeliness and feasibility were evaluated. Ramboll, Denmark, through joint design with wind turbine manufacturers, uses interactive iterative design to complete the foundation design. It cooperates with the Danish Technical University (DTU) and the National Renewable Energy Center to study the integrated optimization of offshore wind turbine support structure, and cooperates with MHI Vestas to successfully develop the first packaged integrated load calculation software SMART Foundation Loads in the offshore wind power industry. The calculation software has the function of early optimization calculation to wind turbine foundation design, which enables the calculation ability of wind turbine-foundation coupling in the front-end design stage of the project. It can be applied to the pre-bidding and post-bidding design of offshore wind turbine projects, which can effectively reduce the pre-design interface, improve the load calculation accuracy, eliminate the structural design redundancy, reduce the project risk and project cost. And the possibility of winning the bid can be improved.

The research and development of offshore wind turbines in China started late. There are few related studies. And no systematic results have been formed. At present, the domestic offshore wind turbine is still based on the traditional design, but some wind turbine manufacturers are moving towards integrated design. Goldwind Technology has built iDO, an integrated platform for control, load and tower-foundation. It is based on the cloud platform. It has cooperated with domestic design institutes in integrated design and has been applied in actual offshore engineering projects. Through technological innovation, the development cost of offshore wind turbines is effectively reduced [4]. Shanghai Electric has constructed the preliminary practice of integrated design based on the database formed by offshore projects. By further introducing design constraints, the preliminary scheme is checked and optimized. And the final conceptual scheme is obtained. The MySE5.5 MW typhoon-resistant floating wind turbine, developed by Mingyang Smart Energy, has been assembled and tested at Mingyang Smart Energy's Yangjiang base. It has become the first offshore floating wind turbine prototype installed and applied in China. In the research and development process of offshore floating large megawatt wind turbines, based on the consideration of harsh environments such as typhoons and extreme ocean waves in the South China Sea, the floating wind turbines are deeply studied and customized. The integrated fully coupled time-domain simulation of wind turbine-floating body-mooring is realized, and a mature fully coupled integrated simulation design platform for floating wind turbine foundation is formed. The China General Certification Center has gradually formed a complete evaluation scheme for the safety and effect evaluation of integrated design. It has carried out the first batch of offshore wind power project certification and offshore integrated design evaluation certification pilots in China. For Huaneng Cangnan No. 4 offshore wind power project, the integrated design evaluation of the offshore wind turbine-support structure under typhoon is following the certification. This is also the first offshore wind turbine integrated design evaluation certificate.

3.2 Reliability Design

The operational conditions for offshore wind power development in our country are notably more complex compared to those in Europe. For instance, the southeastern coastal regions frequently experience typhoons, while floating ice is common on the sea surface in northern areas during winter. Additionally, corrosive factors such as high salinity and fog significantly contribute to the increased failure rates of wind turbine units. In comparison to onshore wind turbines, offshore wind turbines present greater challenges in terms of accessibility for operation and maintenance. Major component failures often necessitate the use of large specialized vessels for repairs and replacements, resulting in elevated maintenance costs and imposing stringent requirements on the reliability of offshore wind power equipment [8].

3.2.1 Overview

Reliability design refers to a mechanical design method aimed at eliminating potential defects and weak links in a product, preventing failures, and ensuring that the mechanical system and its components meet the given reliability requirements. It includes the prediction, allocation, technical design, and evaluation of product reliability.

The basic principles of reliability design are as follows:

- There should be clear reliability indicators and a reliability assessment plan.
- Performance, reliability, cost, time, and other factors should be balanced to make the best design solution.
- Reliability factors should be fully considered throughout the functional design process while meeting basic functions [9].
- The design solution should be simplified as much as possible, with a reduction in the number and types of components and equipment. Immature new systems and components should be avoided, and existing standardized components and mature technologies should be used whenever possible.
- The applicability of functional components should be considered, and modular structures can be adopted to enhance maintainability.
- Design should address failure modes (manifestations of system, component, or device failures) to eliminate or control potential failure modes throughout the product's lifecycle.
- When designing for maintainability, the number of tools required for repairing the entire product should be minimized, and the use of specialized tools should be avoided as much as possible [10].

The purpose of reliability design is to achieve a comprehensive balance between reliability and cost-effectiveness by considering the performance, reliability, cost, and design factors of the product. It aims to prolong the service life of mechanical equipment and reduce the operational and maintenance costs of wind power turbines [11]. Relying solely on experience during the product design process without paying attention to performance requirements and reliability often leads to quality issues during the testing phase, requiring further improvements and extending the product development cycle. However, conducting reliability design, scientific reasoning, and rigorous argumentation in advance can significantly shorten the product development cycle [12]. From a long-term perspective, reliability design adopts measures during the design phase at a relatively low cost, resulting in excellent effects and reduced time consumption. It is an economically viable design approach.

Common reliability design methods include safety margin design, redundancy design, derating design, thermal design, simplified design, electromagnetic compatibility design, etc.

1. **Safety Margin Design**

Safety margin design is the establishment of a larger safety factor for important mechanical components. Although it is essentially a measure of reliability, it is not infinite and is always subject to other constraints. The task of safety margin design lies in determining a reasonable margin under the limitations of other factors such as quality, volume, and cost. Wind turbines, due to factors such as storms, gusts, and waves, may experience emergency conditions where the loads exceed the average value. Having a certain safety margin can prevent damage from occurring [13].

The general process of safety margin design is as follows:

- Decompose the complex product into several stages and identify the weak links;
- Determine the generalized reliability safety margin design characteristic quantity and define the reliability of the weak links;
- Seek the critical central value of the design characteristic quantity through testing;
- Identify the distribution pattern of the critical central value and determine the distribution parameters;
- List the generalized safety margin design equation and determine the design value of the characteristic quantity;
- Calculate the reliability index and the design value of the characteristic quantity corresponding to each stage [14].

2. **Redundancy Design**

Redundancy design involves incorporating backup hardware or structures into the system, either actively participating in its operation or remaining in a state of readiness. This ensures that the system can continue performing its designated functions and maintain uninterrupted normal operation, even in the event of a failure. Redundancy design can be implemented at various levels, ranging from components, parts, and modules to entire subsystems, and is an effective measure to enhance equipment reliability. It is also referred to as margin design or reserve design. When the reliability requirements for a certain component are particularly high but cannot be met by the current technological capabilities—such as through derating design, simplified design, or other reliability design methods—or when the cost of improving component reliability exceeds that of redundant configuration, redundancy technology may become the only or most favorable design solution. This approach can combine less reliable individual components into an overall system with higher reliability, and is commonly applied in electronic products. As mechanical systems become increasingly complex and reliability demands rise, redundancy design can be employed by weighing trade-offs between cost, weight, and reliability. However, redundancy also increases system complexity, which raises the question of whether redundancy is necessary, and if so, what type of redundancy is most appropriate. The decision depends on the specific devices and their criticality, as well as whether the benefits outweigh the costs. In general, an effective redundancy design can both enhance system reliability and reduce costs [15].

3. **Derating Design**

Derating design is a design method that diminishes the operational stress of electronic components to a level beneath their rated stress, with the objective of augmenting component durability, postponing parameter deterioration, lowering failure rates, and prolonging the lifespan of the electronic components. The failure rate of electronic components is highly sensitive to electrical stress and temperature stress, making derating design one of the most commonly used methods in reliability design for electronic products. When implementing derating design, the greater the reduction in stress, the higher the required performance of the electronic components, and consequently, the higher the cost. Therefore, a comprehensive consideration is necessary. Each type of electronic component has an optimal derating range. For some, once the stress is reduced to a certain level, it remains constant. For example, some capacitors may exhibit an open circuit at low voltages, where further derating not only fails to reduce the failure rate but actually increases it. Therefore, the following points should be considered during practical derating design:

- Standard-recommended derating values should not be applied rigidly; appropriate adjustments should be made based on the specific characteristics of the product.
- It should be noted that the parameters of certain electronic components should not be arbitrarily reduced.
- Generally, less applied stress on electronic components enhances dependability; however, this is not universally applicable.
- Derating should not be used as a compensatory measure to address the issues associated with low-quality electronic components. Such components should be used with caution [16].

4. **Thermal Design**

Thermal design involves employing appropriate and reliable methods to control the temperature of all electronic components within a product, ensuring that the working environment conditions do not exceed the maximum temperature required for stable operation. This ensures the safety of the product during normal operation and its reliability over long-term use. The three common strategies in thermal design are power reduction, heat conduction, and layout. Power reduction prevents heat generation; heat conduction involves dissipating heat promptly to avoid affecting the operation of the equipment; and layout involves isolating heat-sensitive electronic components through appropriate measures. Power reduction is the most fundamental and original solution. Derating and low-power design are the two main approaches [17].

The basic requirements for thermal design are as follows:

- It must be carried out simultaneously with electrical and structural design, considering each aspect in conjunction. When conflicts arise, a balanced analysis should be conducted without compromising electrical performance. It must meet reliability requirements and minimize the lifecycle costs of the equipment.
- It should meet the thermal environment requirements of the product throughout its lifespan.
- It should meet the product's reliability requirements to ensure that all electronic components within the equipment can function normally over the long term under the specified thermal conditions.
- The selection of parameters, installation location, and methodology for each electronic component must adhere to thermal dissipation standards.
- The simplest and most effective cooling method should be selected based on factors such as heat generation, ambient temperature, allowable operating temperature, reliability requirements, as well as considerations of size, weight, cost, and safety.
- Appropriate design margins should be considered to avoid increases in heat dissipation and flow resistance caused by changes in operating conditions during use.
- Economic efficiency should be considered, ensuring that the structure is simple, reliable, compact, and cost-effective, while still meeting the necessary heat dissipation requirements.

5. **Simplified Design**

In the design process, the size accuracy, positional tolerances, and structure of the product should be simplified as much as possible while still meeting performance requirements. This approach facilitates manufacturing, assembly, and maintenance. Mechanical products should be analyzed based on reliability models, as most belong to series systems. Therefore, the key principle for enhancing reliability is to simplify the design while fulfilling the required functions. From selecting reliable components to reducing their number and simplifying the structure, minimizing the number of components is essential. Simplicity and reliability are fundamental principles of reliability design and the most effective methods for reducing failures and improving overall reliability.

6. **Electromagnetic Compatibility Design**

When performing its intended tasks, the electrical system must maintain its performance across various electromagnetic environments without degradation, and its parameters must remain within allowable limits. It must operate in a coordinated and effective manner by enhancing resistance to electromagnetic interference (EMI) and reducing emitted EMI, thereby preventing failures caused by interference and improving reliability. Electromagnetic compatibility design typically focuses on suppressing interference sources and cutting off interference transmission paths. Electromagnetic interference primarily originates from three sources: functional interference sources, non-functional interference sources, and natural interference sources. Based on transmission paths, interference sources can be classified as conducted or radiated, and based on frequency bands, they can be categorized as narrowband or broadband interference sources [18].

The basic principles of electromagnetic compatibility design are as follows:

- At high frequencies, compared to leaded capacitors, priority should be given to using feedthrough capacitors or base capacitors with smaller lead inductance for filtering.
- Large inductance has large parasitic capacitance. To improve insertion loss in the low-frequency range, avoid using single-stage filters. Instead, use multi-stage filters composed of several small inductors.
- Use shielded relays as much as possible and ensure the shield casing is grounded. The interconnection signal lines inside the equipment must use shielded cables to prevent interference coupling.
- To ensure that each shield is connected to its respective pin, select connectors with a sufficient number of pins.

3.2.2 Reliability Design of Mechanical Components

A mechanical system refers to a complex entity composed of several mechanical components that are organically combined to perform a specific function. The reliability of a mechanical system depends on two factors: the first is the inherent reliability of the mechanical components themselves, that is, the ability of each component within the system to perform its required function; the second is the manner in which these components are assembled into the system. The reliability of the mechanical system varies based on the nature of the interconnections between components and the method of assembly [19].

The purpose of reliability design in mechanical systems is to achieve an optimal design through the comprehensive coordination of technical performance, quality, cost, time, and other factors, while meeting specified reliability indicators and fulfilling intended functions. Alternatively, it aims to attain high reliability within the constraints of performance, quality, cost, time, and other requirements.

1. **Safety Margin Design for Mechanical Components**

The operational environment of offshore wind turbine units is harsh, with continuously varying loads caused by factors such as typhoons, ocean waves, and thunderstorms. In more severe cases, these loads can exceed the design limits. However, if the design incorporates a certain safety margin, potential damage can be avoided. Safety margin design is primarily applied to mechanical components in wind turbine units that bear more complex and substantial loads, such as the main shaft, hub, and tower. The safety margin values may vary depending on the specific engineering project and the type of foundation. In the construction of wind farms, due to the influence of various complex factors, the foundation must have sufficient safety margins to ensure it is not compromised under extreme load conditions. Therefore, when designing the foundation, the safety margin should be determined based on the most adverse load conditions [20].

2. **Anti-loosening Design for Fasteners**

(1) Causes of Fastener Loosening

Threads are a mechanical structural element used for connection and transmission, offering advantages such as high connection strength, compact structure, ease of assembly and disassembly, and reliable connection. As a result, they have become the most widely used connection method. In some critical areas, the failure of threaded fastener connections can result in immeasurable losses. Due to the limitations inherent in the design of threaded structures, loosening is one of the primary failure modes of threaded fasteners. Therefore, effective preventive measures must be taken, leading to the development of various anti-loosening threaded fasteners. Under static load conditions, threaded fasteners only bear axial loads. Due to the influence of the thread helix angle, the torque required to tighten the nut differs from the torque required to loosen it. Typically, the loosening torque is about 80% of the tightening torque. In the absence of additional torque, the connection will not loosen. Variable loads, vibration, and impact are the primary factors contributing to the loosening of threaded fasteners. Most threaded fasteners are used in environments with vibration or impact. Due to the inertia of the fasteners and the interaction with the connected components, the friction coefficient between the thread pair and the nut bearing surface decreases sharply, and in some cases, the friction resistance may momentarily disappear. This disrupts the original balance of forces, preventing the thread pair from maintaining the self-locking condition, and results in slight relative sliding [21].

(2) There are three basic types of anti-loosening mechanisms for threaded fasteners

- Non-removable anti-loosening: This method involves converting a detachable threaded connection into a non-removable one by welding, bonding, or punch riveting. It is a highly reliable traditional anti-loosening method. However, the downside is that the fasteners cannot be reused, and the operation is cumbersome. This method is typically used in critical situations where high anti-loosening reliability is required and disassembly is not necessary. An example of punch riveting on the bolt head and nut end surface is shown in Fig. 3.3.
- Anti-loosening of mechanical fasteners: This method involves using mechanical fasteners to secure and lock threaded fasteners to either the fastened components or other threaded fasteners, thereby preventing loosening. The advantage of this approach is its reliable anti-loosening performance, which generally depends on the static or fatigue strength of the mechanical fasteners. However, the disadvantages include increased mass of the fasteners, more complex manufacturing and installation, the inability to allow for adjustable installation, and higher costs. An example of a locking wire is shown in Fig. 3.4.

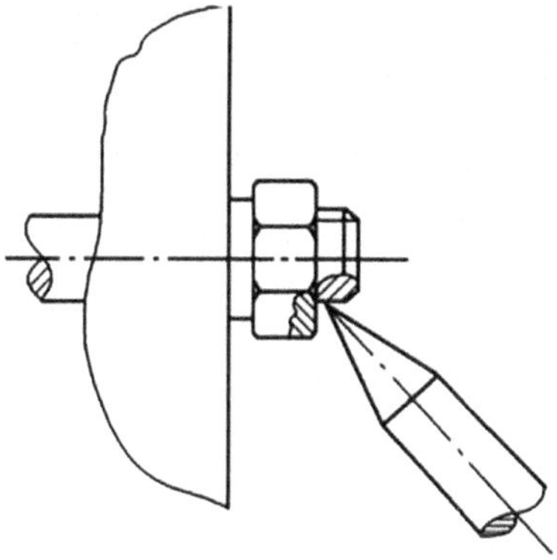

Fig. 3.3 Punch riveting on the bolt head and the nut end surface

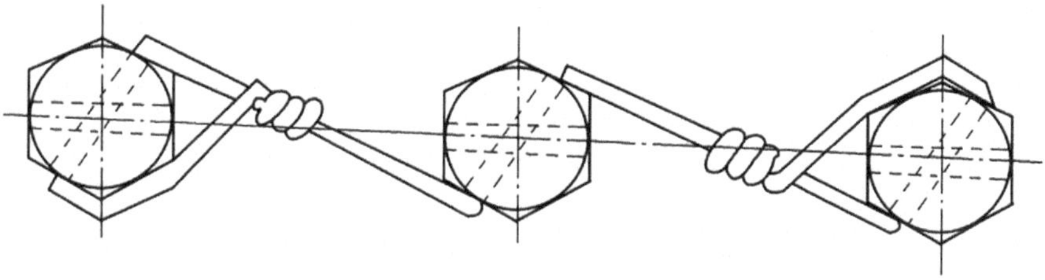

Fig. 3.4 The locking wire

Increasing friction for anti-loosening: This method achieves anti-loosening by increasing the friction between the threads or between the bolt (screw) head and the nut end surface, or by increasing the friction in both areas simultaneously. While this method is generally less reliable than the previous two, its greatest advantage is that it is not restricted by space, allows for multiple repeated assembly and disassembly, and supports mechanized assembly. The anti-loosening reliability of certain fasteners (such as nylon insert lock nuts and all-metal lock nuts) has reached a very high level. An example of a spring washer is shown in Fig. 3.5.

(3) Basic Principles of Anti-Loosening Design

- Give priority to using bolt and nut connections, and avoid using bolt-to-internal-thread connections whenever possible. If unavoidable, the internal thread should be designed on components that are easier to replace or have lower cost.
- The fastening bolts should be designed to ensure they are not subjected to shear forces. For critical connections that may be exposed to shear forces, measures such as using bolts with pins or reamed bolts should be implemented.
- Critical load-bearing threaded connections should have no fewer than two anti-loosening mechanisms, with at least one being a mechanical anti-loosening

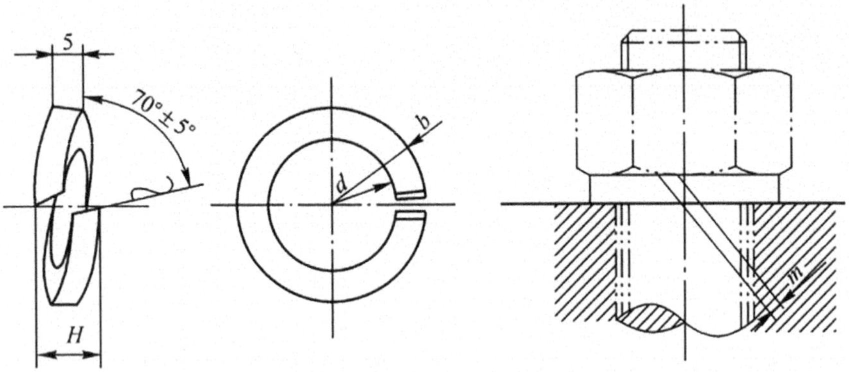

Fig. 3.5 Spring washer

method other than spring washers or tightening torque. Adhesive and spring washers cannot be considered reliable anti-loosening measures.
- When designing bolt assemblies, sufficient spacing must be ensured to allow for proper installation.
- In principle, threaded fasteners should not be designed to be tightened from below upwards [22].

3.2.3 Reliability Design of Gearboxes

The gearbox is a critical component of wind turbine units. Among mechanical failures in wind turbines, gearbox failures represent a significant proportion and have severe consequences. Therefore, the reliability design of the gearbox is of great importance in ensuring the safe and reliable operation of wind turbines. Table 3.3 shows the classification of common gearbox failures [22].

Table 3.3 Common fault classification of gear boxes

Common fault location	Fault type
Gear	Gear broken, tooth surface erosion, glue
Transmission shaft	Broken shaft
Bearing	Bearing normal high temperature and abnormal installation position
Lubrication	Oil temperature is too high, the oil supply is insufficient, and the lubricating oil viscosity is reduced
Box binding surface	Oil leakage, low inlet pressure of lubricating oil, abnormal oil level

At present, extensive research has been conducted both domestically and internationally on gearbox reliability. The most widely applied method is the system reliability approach commonly used in the aerospace, nuclear energy, and offshore industries: Failure Mode, Effects, and Criticality Analysis (FMECA). FMECA consists of two parts: Failure Mode and Effects Analysis (FMEA) and Criticality Analysis (CA). CA analysis can only be performed based on the foundation of FMEA analysis.

The steps of FMECA are as follows [23].

(1) Define the scope of analysis

Determine the scope of the analysis based on the system structure and the objectives of the analysis.

(2) Structural and functional analysis of the system

Create a block diagram of the structure, and based on this diagram, complete the structural and functional analysis.

(3) Failure mode analysis

Throughout the system's lifecycle, analysts typically obtain failure modes through statistical analysis, predictive analysis, testing, and empirical summarization. For common components, standards and manuals can be referenced. Common failure modes and their corresponding codes are shown in Table 3.4.

(4) Failure mode analysis

The causes of failure can be considered from two aspects: direct causes and indirect causes. Direct causes are those that arise from internal factors, while indirect

Table 3.4 Common fault modes and codes

Fault-pattern	Code	Fault-pattern	Code	Fault-pattern	Code
Bend deformation	F01	Loosen	F09	Bearing fracture	F17
Flaw	F02	Surface damage	F10	The holder injury	F18
Break	F03	Gear ring deformation	F11	Motor pump failure	F19
Dentistry wear and tear	F04	Bearing wear and tear	F12	Filter fault	F20
Tooth surface plastic deformation	F05	Bearing plastic deformation	F13	Thermal exchanger failure	F21
Wheel teeth broken	F06	Rustiness	F14	Line, check valve fault	F22
Tooth surface scuffing	F07	Fatigue flake	F15		
Pastry etching	F08	Fire burn	F16		

causes generally refer to external factors, such as failures of other components, environmental conditions, or human factors.

(5) Failure effects analysis

The effects of failures can be classified into three levels, as shown in Table 3.5 [24].

(6) Severity

Severity refers to the level of seriousness of the impact a failure has on the system after it occurs. The severity levels are listed in Table 3.6 [25].

(7) Failure detection method analysis

Failure detection method analysis is conducted based on individual failure modes. Common failure detection methods include visual inspection, sensor-based detection, alarm or fault indication devices, and system operational data analysis. Failure detection methods are classified into pre-failure detection and post-failure detection, both of which provide important references for the design, manufacturing, and maintenance of wind turbines. For large wind turbines, given the design lifespan requirement of 20 years, the goal is typically to avoid failures or maintain an extremely low failure rate, rather than relying on post-failure detection methods. Therefore, the focus should be on pre-failure detection in the areas of reliability design, manufacturing, and assembly.

(8) Analysis of design improvements and compensatory measures

Table 3.5 Fault impact

Name	Definition
Local impact	Impact of component failure on peer components
The impact of the upper level	Effect of component failure on the previous level
The final impact	The final effect of a component failure on the convention factor

Table 3.6 Severity grade

Criteria grade	Impact severity
Level 1	Severe damage, wind turbine shutdown
Level 2	The gear box system has moderate damage, obvious vibration and abnormal operation
Level 3	Gear box system has mild damage, slight vibration and accelerated wear
Level 4	The gear box system is basically free of damage and normal wear

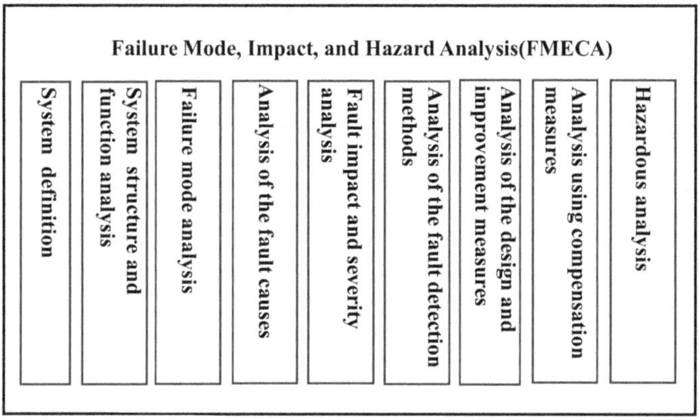

Fig. 3.6 FMECA flow

To improve the overall reliability of the gearbox, it is generally necessary to propose possible improvements and compensatory measures for the causes and effects of each component's failure mode. Such measures may include the use of redundant equipment, safety and protection devices, and process improvements in the event of equipment failure. The FMECA process is shown in Fig. 3.6. FMECA begins by defining the scope of analysis, followed by an analysis of the system's structure and functions, and concludes with the completion of the FMECA report.

3.2.4 Reliability Design of Generators

A wind turbine unit consists of hundreds of components, with key components including the generator, gearbox, bearings, and blades, all of which have high maintenance costs. Statistics show that generator failures are one of the primary causes of wind turbine downtime, with long response times, often leading to extended shutdown periods. Generator maintenance costs account for 10% of the total maintenance costs of wind turbines [26].

Table 3.7 shows the distribution of failures in generator components [27]. Figure 3.7 presents the failure rates of generator components across different capacities. It is evident that, compared to generators of the same specifications used in other industries, the reliability of generators in wind turbine units is relatively lower. The failure locations of generators indicate that the failure points in wind turbine generators are no different from those in generators used in other industries, with the main failure areas being bearings, slip rings, and brushes.

Among generator failures, rear bearing temperature faults account for as much as 90%, making it the most frequent and impactful fault affecting the generator. Rear bearing temperature faults in the generator directly affect its normal operation, causing irreversible damage to the generator. This, in turn, impacts the normal

Table 3.7 Distribution of generator fault sub-components

Survey type	The IEEE survey of large rotors	Corresponding to the investigation of electric motors used for public and facilities	Investigation of the motor of facilities for offshore petroleum, motor in the industry	Investigation of wind turbine generators		
Industrial scope	Ensemble	Communal facilities	Offshore petrochemical industry	Wind power generation		
Motor type, rated power and voltage level	Higher than 150 kW motor, usually medium voltage, high voltage cage induction motor	Higher than 75 kW motor, usually medium voltage, high voltage cage induction motor	Higher than 11 kW motor, usually medium voltage, high voltage cage induction motor	Less than 1 MW wind turbine, low voltage, more than 95% of the winding rotor, motor, but the rotor, the voltage is usually controlled by electricity, sub, but not, is a collector ring with external electronic equipment	1–2 MW wind turbine, low voltage, most of the doubly-fed induction generator	More than 2 MW wind turbines, low voltage, most of the doubly-fed induction generators
Failure under investigation, number of generators	360	1474	1637	196	507	297
Subcomponent bearing (%)	41	41	42	21	70	58
Coolant passage (%)	–	–	–	–	2	–
The stator groove wedge (%)	–	–	–	–	–	14
Certain-related faults (%)	37	41	13	24	3	15

(continued)

Table 3.7 (continued)

Rotor-related fault (%)	10	9	8	50	4	4
Collecting current ring or collector ring (%)	–	–	–	1%	16	4
Rotor wire (%)	–	–	–	–	1	4
Other (%)	12	14	37	4	4	1
Total (%)	100	100	100	100	100	100

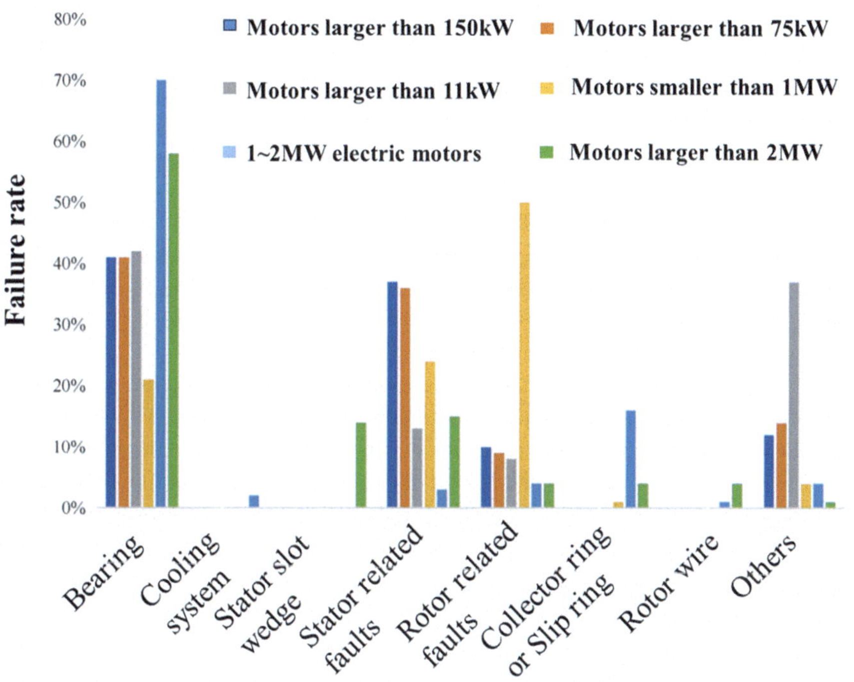

Fig. 3.7 Comparison of failure rates of various subcomponents in generators of different power ratings [27]

functioning of the wind turbine unit, leading to a reduction in its power generation efficiency. Therefore, predicting the future temperature of the generator's rear bearing and understanding its trend over time can provide on-site engineers with guidance in adopting appropriate maintenance strategies. This can reduce the frequency of generator failures, minimize damage to the generator system, improve operational reliability, and extend the generator's service life [26].

In conclusion, if we can analyze the data patterns based on the state data from the wind power monitoring system, conduct real-time assessment and monitoring of the generator's health status, and provide an effective fault prediction, it will provide scientific and rational guidance for wind power engineers to develop maintenance strategies. This is of great significance in reducing the frequency of failures and extending the lifespan of the generator.

3.2.5 *Reliability Design of Converters*

The converter is a core component of wind turbines, capable of influencing the output power and grid stability of the wind turbine. According to statistical data [28], the failure rate of converters in wind turbines is higher compared to other industrial sectors. During the operation of wind turbines, wind speed and ambient temperature fluctuate randomly, causing continuous changes in the power processed and converted by the converter. The operating conditions frequently switch, subjecting the converter to repeated high electrical, thermal, and mechanical stresses, which accelerate its wear and ultimately reduce its reliability [26]. Regarding the reliability of internal components in converters, data collected from several decades of converter failure rates at multiple European wind farms indicate that the main failure components are the direct current (DC) side capacitors, power devices, and drive systems, as shown in Fig. 3.8 [28]. The reliability of components such as the DC side capacitors and power devices directly affects the reliable operation of both the wind turbine and the power grid.

With the development of wind power technology and the increase in wind turbine capacity, the power system's requirements for the reliability of wind power equipment are becoming increasingly stringent. As shown in Fig. 3.8, more than 20% of converter failures are caused by the failure of DC side capacitors. As one of the most vulnerable components in the converter, DC side capacitors should receive sufficient

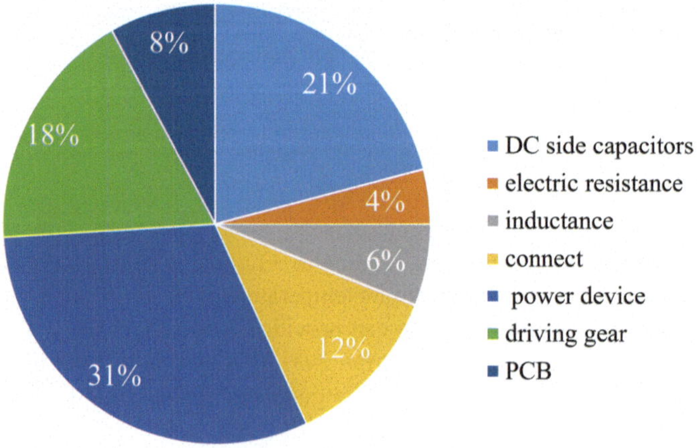

Fig. 3.8 Failure rate distribution of components in converters

attention. This is because DC side capacitors are often used as energy storage devices in converters, primarily serving to absorb the high-amplitude ripple current drawn by the inverter unit from the DC side and to maintain DC voltage stability. They balance the power differences in the converter system caused by fluctuations in wind speed and ambient temperature during the operation of wind turbines.

At present, commercialized high-power offshore wind turbine converter systems are mainly divided into full-power converter systems and partial-power converter systems.

Full-power converter systems are widely used in high-power offshore wind turbines due to their wider speed regulation range, elimination of generator slip rings, and strong grid compatibility. Partial-power converter systems are primarily used in doubly-fed wind turbines.

Compared to the partial-power converter system of doubly-fed induction generators, the full-power converter system of permanent magnet synchronous generators has a lower failure rate and higher availability.

Research by the German Wind Energy Institute and the UK Renewable Energy Association indicates that the converter is one of the three most critical components in offshore wind turbines, accounting for 17–25% of total failures and 3–8% of total operation and maintenance costs. Among these, failures that can be remotely reset result in a 0.1% loss in availability, while failures that require personnel to enter the wind turbine for maintenance—due to the long transit times and limited accessibility of offshore wind operations—can result in availability losses of up to 1.1% [8].

Improving converter reliability and reducing operation and maintenance costs are urgent needs for the offshore wind power industry: One approach is to enhance the reliability of the converter itself, which is relatively complex and requires high costs to achieve. Another approach is to achieve high energy availability and low operation and maintenance costs for wind turbines through parallel fault-tolerant operation of converters.

3.2.6 Prospect

Currently, domestic wind power engineers in China lack sufficient operational experience with offshore wind turbines and do not have comprehensive engineering and operational data to support reliability design. In current designs, offshore wind turbines adopt more conservative safety factors for various structural components compared to onshore wind turbines to ensure operational reliability. How to balance the reliability, safety, and cost-effectiveness of wind turbines, and select safety factors that are suitable for China's offshore environment and wind resource characteristics, still requires further exploration.

The key approach to enhancing the competitiveness of the wind power industry is reducing the LCOE. In 2020, although onshore wind power had achieved price parity with coal-fired power, the LCOE for offshore wind power remained relatively high. This was primarily constrained by factors such as the construction, operation, and maintenance of offshore wind turbines, leading to a slow reduction in the LCOE.

The complex and harsh offshore environment keeps the failure rate of wind turbines consistently high. The corrosive effects of high salt and high humidity air environments pose many challenges to reliability design. The reliability of wind turbines can be improved through various technical measures. The following briefly introduces two conventional technical measures that still hold great potential:

- Multi-objective control technology: The core task of wind power generation is to achieve optimal power output while ensuring the reliability and safety of wind turbines. The load on wind turbine components is closely related to the design of control algorithms. Currently, commonly used power-control-oriented algorithms give limited consideration to factors such as improving the load on components. By improving existing control algorithms to achieve multi-objective control of wind turbine operations—both regulating power output and reducing component loads while extending the overall lifespan of the turbine—this has become one of the key future research directions.
- Adoption of new transmission chains: The choice of transmission chain for wind turbines is crucial for cost reduction, especially for gearbox manufacturing companies. The selection of transmission chains for wind turbines can be divided into two main types: direct drive and geared with a gearbox. Each has its advantages and disadvantages. Direct drive wind turbines have low failure rates and low maintenance costs, but they have high manufacturing costs, large size, and permanent magnet materials that are prone to demagnetization under high temperatures, shocks, and vibrations, with poor on-site maintainability. Geared wind turbines, although lightweight and with low manufacturing costs, have higher failure rates. Currently, low gear ratio gearbox semi-direct drive wind turbines have shown certain advantages and have gradually matured, promising to play an important role in the future.

3.3 Typhoon-Resistant Design

3.3.1 Basic Characteristics of Typhoons

Typhoons are low-pressure cyclones formed in tropical or subtropical regions with maximum wind speeds at the center reaching or exceeding 12 on the Beaufort scale. They have the following characteristics:

- Widespread area of influence: A typical typhoon comprises a wind field, rainstorm area, and an eye, with a diameter ranging from 500 to 1000 km. The most destructive region, the cloud wall around the eye, has a 10–20 km width.

- High average wind speed and rapid turbulence: The turbulence intensity at the typhoon center can reach 0.6 to 0.9 (usually < 0.1 without a typhoon). When the typhoon center approaches, the turbulence intensity changes asynchronously between the lower and upper layers, with a time difference of several tens of minutes.
- High rate of wind direction change: When a typhoon passes, the wind direction at a measuring point can change by more than 180° within a few hours, which places higher reliability requirements on the yaw system of wind turbines.
- Rapid wind shear: Typhoon wind speeds can reach 3 to 5 times or even more than the rated wind speed of wind turbines. Under high wind speeds, large wind shear will significantly increase the vibration response amplitude of wind turbines.
- Long duration, accompanied by typhoon waves: The duration of a typhoon's impact on a specific location is generally 12 to 24 h, accompanied by typhoon waves over 6 m high. The combined action of typhoons and giant waves can easily cause structural damage and fatigue failure of wind turbines. Additionally, phenomena such as excessive seabed scouring, shear failure, and seabed liquefaction are prone to occur [29].
- Extreme Wind Speed: Super extreme wind speed is a prominent feature of typhoons. On September 2, 2003, the 13th typhoon "Dujuan" landed in Shanwei. When it landed, the maximum wind near the typhoon center reached Lv.12, and the extreme wind speed measured by a wind farm near the typhoon center was 57 m/s. On May 18, 2006, the No. 1 strong typhoon "Pearl" crossed Nanao Island and landed in Chenghai, Guangdong Province. When it landed, the maximum wind near the typhoon center was Lv.12, and the instantaneous wind speed was as high as 56.5 m/s measured by a wind farm in Nanao. On August 10, 2006, the 8th super typhoon "Sang Mei" landed on the coast of Cangnan, Zhejiang Province. The maximum wind near the center of the typhoon was Lv.17 (60 m/s), and the central pressure was 92 kPa. The extreme wind speed observed in Xiaguan, Cangnan, Zhejiang Province, is 68.0 m/s, and the extreme wind speed observed in Hepalma Rock, Fuding, Fujian Province, is 75.8 m/s [29].

3.3.2 Wind Turbine Damage Caused by Typhoon

In September 2003, typhoon "Dujuan" made landfall in Shanwei, causing damage to 13 wind turbines at the Honghai Bay wind farm. In 2006, typhoon "Sangmei" hit the wind farm in Cangnan, Zhejiang Province, resulting in the instant collapse of five wind turbines and broken blades, with economic losses of up to 70 million yuan [31]. In 2017, Typhoon Dove caused a large number of broken wind turbine blades and the destructive collapse of several wind turbines at an onshore wind farm in the Pearl River Delta, resulting in serious economic losses and even the uprooting of some wind turbines. At the same time, ships near an offshore wind farm in the Pearl River Estuary were affected by the typhoon and struck the wind turbine foundation trusses, causing them to be severely deformed. Some of the submarine cables were damaged by the ship's anchor gear during the typhoon, and the submarine cables required a long repair period, which had a serious impact on the repair schedule and resulted in huge economic losses [32]. Typhoons can cause severe damage to wind turbine blades, braking systems, nacelles, wind measurement devices, towers and infrastructure, with the most frequent cases being overall overturning, tower failure and blade damage, as shown in Figs. 3.9, 3.10 and 3.11 respectively [31].

Fig. 3.9 Overall capsize of wind turbine [31]

Fig. 3.10 Wind turbine tower failure [31]

Fig. 3.11 Wind turbine blade destruction [31]

3.3.3 *Typhoon-Resistant Design Techniques*

Typhoon-resistant design of offshore wind turbines refers to a scientific and reasonable typhoon-resistant design concept based on a full understanding of the basic characteristics of typhoons and the failure modes of offshore wind turbines under this condition, with effective typhoon-resistant measures proposed to ensure that offshore wind turbines achieve a "two-stage" typhoon-resistant design: In typhoons where the maximum wind speed is less than the design wind speed, the main structures and components are not damaged; in typhoons where the maximum wind speed exceeds the design wind speed, damage is kept within the expected range and no overturning damage occurs, and after the typhoon, the offshore wind farm can be quickly repaired and put into operation.

The typhoon protection design should ensure that the wind turbine can actively yaw against the wind: strengthen the design of the load safety chain to ensure that all components of the wind turbine can operate normally; strengthen the design of mass dampers in areas where typhoons occur frequently to reduce the vibration of the wind turbine; strengthen the typhoon protection design of the nacelle cover so that it is intact during typhoons to protect the internal components; strengthen the fixation of the wind speed and direction meter to ensure that it can operate normally during typhoons.

1. **Typhoon-Resistant Design of the Blades**

When wind farms are hit by strong typhoons, the typhoon can cause serious damage to the blades of the wind turbines such as broken blades, collapsed towers and so on. Under the influence of the typhoon, the blades are usually subjected to bending, torsional and shear forces. Under the action of bending, torque and shear force, a complex cross-stress force is formed in the blade section, and cracks or even destructive fractures occur when a certain fatigue state is reached. Therefore, in order to avoid greater losses, it is necessary to further strengthen the local defect detection in the production process of blades and improve the ability to resist typhoons. When the wind speed exceeds the typhoon resistance limit wind speed of the wind turbine, the blade is allowed to break in a certain position to reduce the wind load of the wind turbine, and protect the tower barrel and other key components; New materials such as carbon fiber are used to improve the ability of blades to resist typhoons or improve weak links. Carbon fiber blades have the following advantages over glass fiber blades [32].

- Quality: the blades are more than 30% lighter than glass fiber blades, exhibiting increased resistance to typhoons, while simultaneously fulfilling stiffness and strength requirements.
- Electricity generation: thin airfoils with higher aerodynamic efficiency can be used to increase wind energy utilization and annual electricity production.
- Costs: Lower overall wind power costs, e.g. lower installation and transport costs, lower requirements for strength and stiffness of wind turbine components, higher overall performance of the wind turbine, lower regular servicing and maintenance costs, Reduced maintenance costs.
- Efficiency: The blade length can be increased without increasing the load on bearings, root fasteners and hubs, allowing more energy to be captured on the same platform.

2. **Typhoon Resistant Design of Tower**

The objective of implementing typhoon-resistant design for wind turbines is to mitigate the risk of overturning damage. Strategies can be employed to enhance the tower's rigidity and taper, as well as reinforce the welding of tower joints. These measures aim to prevent catastrophic damage caused by tower breakage during typhoons. Additionally, the wind turbine can be designed to orient itself in the downwind direction, while streamlining the nacelle to reduce wind resistance and enhance typhoon resistance. In the realm of large-scale offshore wind turbine design, stability concerns have become increasingly prominent due to the utilization of lightweight and high-strength materials in component construction. Consequently, flexural analysis plays a particularly significant role. The tower barrel of offshore wind turbine should fully consider the buckling condition under the influence of typhoon, select high-strength materials, or increase the height of jacket and base platform, reduce the height of tower barrel, and reduce the swing amplitude of the engine room.

Studies have shown that the wind loads acting on reinforced concrete towers are significantly lower than those on steel towers for the same installed capacity [33, 34]; Taking into account the situation in various countries, the damping ratio of steel structure is generally 0.01–0.02, and the damping ratio of reinforced concrete structure is 0.03–0.08. Obviously, the damping ratio of reinforced concrete structure is much higher than that of steel structure, and the choice of reinforced concrete structure with higher damping ratio can withstand greater wind load, which is conducive to typhoon protection. Corrosion-resistant, inexpensive and heavy reinforced concrete towers are now being widely used in wind farms.

Wind turbine bending moments and horizontal loads are large, especially for offshore wind turbines. Reinforced concrete towers have a large self-weight load,

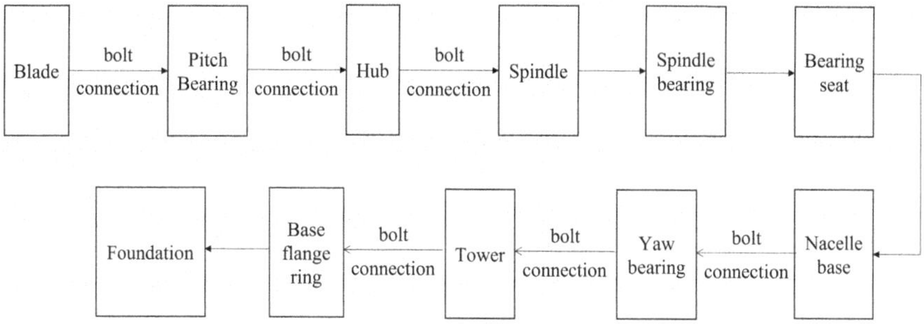

Fig. 3.12 Load transfer safety chains [13]

which is useful for the overall structural overturning of offshore wind turbines and for controlling the base dislocation area.

3. **Load Transfer Safety Chain Enhancement Design**

The load of a wind turbine forms a transfer chain, which must be secured to ensure the wind turbine's strength against typhoons. This chain is known as the load transfer safety chain. Any failure in this chain to meet the requirements of the typhoon design will result in significant damage in the event of a large typhoon. The load transfer safety chain in this case is, of course, different from a normal safety chain, which is on the main PLC and is independent of the control system (hardware implementation). The load transfer safety chain is a closed-circuit chain of safety-related sensors.

The load transfer safety chain includes the blades, pitch bearings, hubs, spindles, spindle bearings, bearing housings, nacelle bases, yaw bearings, towers, foundation flange rings and foundations. Of these, the blade and the pitch bearing, the pitch bearing and the hub, the hub and the spindle, the bearing housing and the nacelle frame, the nacelle frame and the yaw bearing, the yaw bearing and the tower, the tower and the foundation flange ring are all bolted together, as shown in Fig. 3.12.

When designing a load transfer safety chain, the loads on the components of the chain are first calculated and then calibrated for static and fatigue strength. If the static and fatigue strength checks do not pass, the load transfer safety chain will be strengthened by means of material replacement, structural modification, size change, etc. [13].

4. **Reinforced Design of the Nacelle Cover**

The nacelle should be of an integral box type, not a top opening or back lifting type. Because typhoons can blow into the nacelle from all directions, open nacelle covers can be blown away by both the negative pressure caused by high velocity airflow and the positive pressure of high winds blowing into the nacelle. By ensuring that the nacelle and tower are not damaged in a typhoon, 80% of the wind turbines are in perfect condition.

Fig. 3.13 Overall reinforced square nacelle cover [35]

(1) Reinforced design for nacelle cover connections

The strength of the assembled nacelle cover should be checked to ensure that it can meet the design requirements for typhoon resistance. If the typhoon-resistant design requirements are not met, the connection of the nacelle cover shall be strengthened: increase the number of bolts, adopt double-row bolt connections and expand the bolt connection area; increase bolt strength, use higher-strength bolts, and increase tensile strength; Increase the thickness of the connecting part of the nacelle cover to increase the tensile strength and resist typhoon damage; use higher-performance nacelle cover materials to improve the stiffness and strength of the nacelle cover as a whole, facilitate the yaw control of the wind turbine, and better realize wind turbines against the wind during a typhoon.

(2) Complete nacelle reinforcement

The overall reinforcement of the nacelle consists of reinforcing the front, middle and rear parts of the upper canopy: the joint between the upper hood and the lower hood, the joint between the left and right parts of the lower hood, and the inside of the left and right parts of the lower hood. The overall reinforcement of the square nacelle cover is shown in Fig. 3.13.

The front ribs are used to prevent the nacelle from being lifted, and the rear ribs are used to reinforce the anemometer. The front and rear ribs are reinforced to balance the force and reduce typhoon damage.

5. **Backup Power**

Typhoons often cause grid outages. In offshore wind farms, the impeller of the wind turbine will not be able to keep the wind in the right direction in the event of a loss of power to the grid. The most undesirable direction of force is the side of the nacelle, which can easily cause the turbine to fly off even if the blades are properly aligned, leading to blade breakage or turbine collapse. All large offshore wind turbines require a large diesel generator as a back-up power source. This can be installed at one end of the wind turbine's collector line to avoid the loss of AC power to the wind during typhoons, which can cause the blades to fly or the nacelle to collapse due to lateral forces. A container can also be added to the wind turbine platform, in which UPS power and batteries can be prepared to provide yaw and variable pitch power for about one day [32].

6. **Structural Vibration Control Technology**

Structural dampers can effectively suppress the vibration of engineering structures and have been widely used in engineering. Tuned mass dampers (TMD) are often used in mechanical damping devices, while tuned liquid dampers (TLD) and tuned spring dampers (TSD) are also used. A simple TMD consists of a mass block (inertial force), a spring (elastic recovery force) and a damping (dissipation of energy), which can effectively reduce the vibration of the main structure. In recent years, scholars at home and abroad have done some simulation studies on the vibration damping performance of TMD systems for wind turbines. The structure of a control system using a TMD to suppress tower vibration is shown in Fig. 3.14 [36]. Figure 3.14(a) and (b) show the TMD of different installation structures. In the damping control, the frequency of the TMD is tuned to a frequency close to the tower vibration frequency in order to absorb the energy generated by the tower vibration. At the same time when the tower is vibrated by external excitation, the mass block moves in the opposite direction to the tower and the support structure in the opposite direction of the mass block exerts a counter force on the tower vibration to stop the tower vibration. The simulation results of offshore wind turbines by Lackner et al. [37] show that the passive control system of TMD can reduce the fatigue load of tower bottom bending moment by 10%. The research of Rodriguez et al. [38] shows that wind turbine loads can be reduced by 20% and tower design quality by 10% by utilizing technologies such as generator torque control, blade pre-bending technology and TMD vibration reduction system.

TMD is also available with a liquid structure. This liquid-structured TMD is called a tuned liquid damper (TLD). The main principle of TLD is that when the

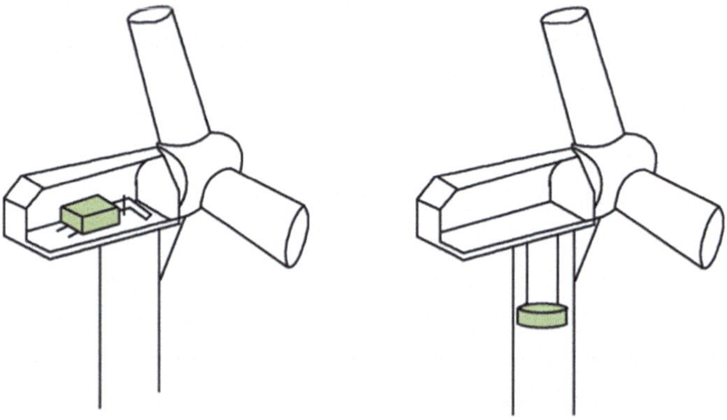

(a) stationary type (b) pendulum model

Fig. 3.14 Structure for suppressing tower vibrations using the TMD's control system

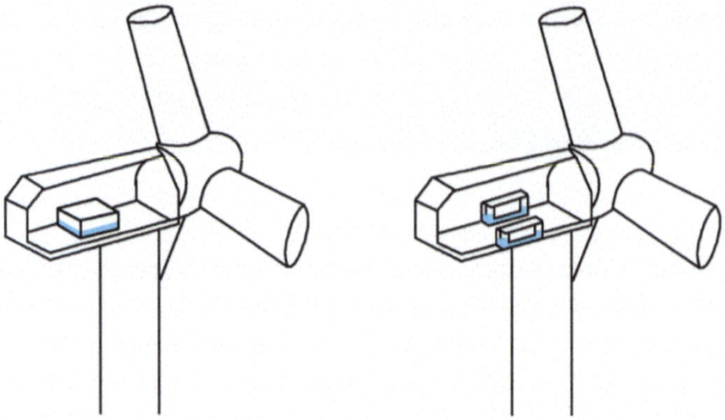

(a) Cuboid structure (b) Column structure

Fig. 3.15 Wind turbines with TLD [36]

tower vibrates, it will be displaced with the tower vibration, causing the liquid to move and using the inertial and viscous forces generated by the shaking and impact of the liquid in the fixed water tank to produce damping, absorbing the energy generated by the tower vibration, thus playing a role in vibration damping. Figure 3.15 shows a wind turbine with a TLD. Compared to TMD, TLD is simple to install, easy to tune, can be used as water supply tanks and do not require any mechanical support components [36].

In the event of a typhoon, the TLD can consume a large amount of energy in a short period of time, thus maintaining the structural safety of the offshore wind turbine and facilitating typhoon-resistant design. In addition, TLD is also effective in controlling the vibration amplitude of offshore wind turbines in the absence of typhoons, extending their working life and increasing their operational stability. It is worth noting that compared to conventional typhoon protection measures, dampers are smaller, lighter, less costly, more effective and more flexible, and in practice are expected to have the effect of making a difference [39].

3.3.4 Typhoon-Resistant Control Strategy

Similar to the "three-level" earthquake resistance, the anti-typhoon design of offshore wind turbines is also roughly divided into "three levels": when the typhoon comes, the wind turbine starts the yaw system; If it is not enough to resist the typhoon, mechanical braking can also be carried out, at this time it is best to keep the blade idling to consume typhoon energy; If it is still not enough to resist the typhoon, you can also consider sacrificing the blades to ensure the safety of the foundation and tower, and then replace the blades or repair the blades after the typhoon has passed [29].

3.3.5 Conclusion

- Typhoon has the basic characteristics of high extreme wind speed, strong non-stationarity, rapid change of wind direction and synchronization with waves. These characteristics are closely related to the typhoon resistance design of offshore wind turbines.
- The common failure modes of offshore wind turbines under typhoon are overall overturning, tower failure, blade failure, etc.
- According to the specific requirements for typhoon resistance in offshore wind turbines, researchers propose optimizing the design of blades, tower structures, transmission systems, and nacelle covers. They recommend installing backup power systems on the wind turbine platform, as well as introducing structural vibration control technology and risk management strategies for the operational modes of offshore wind platforms. These measures can effectively mitigate typhoon-related damage to offshore wind turbines. Given the complex meteorological conditions associated with typhoons, it is essential to continuously investigate the challenges of typhoon resistance in offshore wind turbines. Additionally, it is important to develop design and control measures that are better suited to the unique conditions of China's coastal regions, ensuring the safe, reliable, and rapid advancement of offshore wind power in the country.

3.4 Anti-corrosion Design

China has a vast coastline and abundant wind energy resources at sea. Due to the significant characteristics of high humidity, high salt spray and long sunshine in the offshore environment, the corrosion rate of offshore wind turbines is significantly higher than that of land wind turbines. How to conduct corrosion-resistant design is one of the technical difficulties encountered in the offshore wind power industry [40].

3.4.1 Overview

The specific definition of corrosion has been evolving with the development of materials. Early on, the definition of corrosion was limited to metal corrosion and was generally defined as "the destruction and deterioration caused by chemical or electrochemical reactions between metals and the surrounding environmental medium."

With the rapid development of non-metallic materials, their corrosion damage has gradually gained attention. Since the 1950s, many scholars in the field of corrosion have tended to broaden the definition of corrosion to include all materials, such as metals, concrete, plastics, wood, and other inorganic or organic non-metallic materials. The definition of corrosion has been expanded to "the destruction and deterioration of materials caused by reactions between the material and the surrounding environmental medium [41]."

Corrosion of materials is closely related to the environment. In many cases, materials not only undergo corrosion but also experience physical and mechanical forces, stress, radiation, electrical currents, as well as biological influences simultaneously. Most of these factors tend to intensify the corrosion process, accelerating the degradation and deterioration of the materials.

Offshore wind power equipment mainly involves metal components, concrete structural components, and composite material components. Below is a brief introduction to the basic principles of corrosion, as a basis for understanding corrosion protection measures.

1. **Metal Corrosion**

(1) Classification

Generally speaking, metal corrosion can be divided into three categories based on the mechanisms: chemical corrosion, physical corrosion, and electrochemical corrosion. The intuitive method for distinguishing between these types is by considering the type of environmental medium in contact with the metal (non-electrolyte, liquid metal, and electrolyte solution).

Chemical corrosion refers to the deterioration that occurs when the metal surface directly undergoes pure chemical reactions with the surrounding environmental medium. Chemical corrosion follows the fundamental principles of multiphase reaction kinetics in pure chemistry. Oxidants react directly with the metal surface atoms, producing corrosion products without the generation of electrical currents. In reality, the conditions for pure chemical corrosion are relatively harsh. Only when metals corrode in anhydrous organic solvents or dry gases can it be considered as chemical corrosion.

Physical corrosion refers to the corrosion caused by metal materials forming alloys or liquid metals penetrating into grain boundaries due to physical dissolution. This causes the metal to lose its original strength, resulting in structural failure [41].

Electrochemical corrosion refers to the damage caused by the electrochemical reaction between the metal surface and an electrolyte solution. Electrochemical corrosion follows the basic rules of electrochemical kinetics. The oxidation–reduction reactions of metal corrosion are divided into relatively independent anodic and cathodic areas. During the corrosion process, conjugate reactions occur through the transfer of electrons and metal ions, generating an electric current due to the directional movement of charged particles. The reaction rate of electrochemical corrosion is generally much higher than that of pure chemical corrosion. In the corrosive environment of offshore wind power, electrochemical corrosion dominates. Therefore, understanding the mechanism of metal electrochemical corrosion is of great importance.

(2) Mechanism and Main Influencing Factors

Seawater is a highly saline electrolyte solution. The marine atmosphere is rich in salt mist, which deposits on metal surfaces and forms hygroscopic sea salt films (such as calcium fluoride and magnesium chloride). The environmental characteristics of the ocean make offshore wind turbines prone to electrochemical corrosion.

① Corrode Primary Battery

Immerse one end of a zinc (Zn) metal strip and a copper (Cu) metal strip into a dilute sulfuric acid solution, and connect the other ends with a wire to form a corrode primary battery, as shown in Fig. 3.16. Through observation, it is found that the zinc metal strip is corroded, hydrogen gas is evolved at the copper metal strip, and there is current in the wire. The chemical reaction equation is:

$$Z_n + H_2SO_4 = Z_nSO_4 + H_2 \uparrow .$$

From the general electrochemical corrosion equation, the following processes can be summarized.

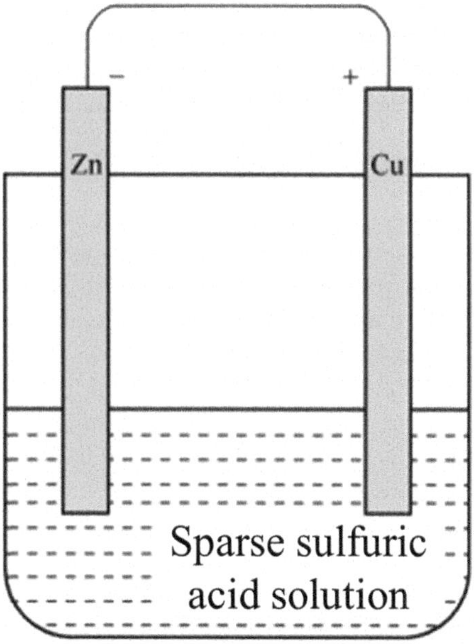

Fig. 3.16 Corrosion of the primary battery

- Anodic process: The metal undergoes dissolution, and the corresponding metal ions enter the electrolyte solution. At the same time, electrons accumulate at the electrode, leading to a metal oxidation reaction. This region is referred to as the anodic corrosion.

$$M_e \rightarrow M_e^{2+} + 2e^- \tag{3.1}$$

- Cathodic process: The oxidative substance accepts electrons from the anode and undergoes a reduction reaction. This region is known as the corroding cathode. In marine environments, the oxidizing agents are typically hydrogen ions (H^+) and dissolved oxygen (O_2).

$$2H^+ + 2e^- \rightarrow H_2 \uparrow \text{ (Acidity)} \tag{3.2}$$

$$2H_2O + O_2 + 4e^- \rightarrow 4OH^- \text{ (Neutral or alkaline)} \tag{3.3}$$

- Current flow: The reactions at the anode and cathode are not completely independent, but are closely related through the migration of electrons and metal ions. In a metal, current flow relies on the movement of electrons from the anode to the cathode. In a solution, it relies on the migration of ions, with cations moving from the anode to the cathode and anions moving from the cathode to the anode. At the interfaces of the anode and cathode, oxidation and reduction reactions occur respectively, facilitating the transfer of electrons. In this way, the entire battery system forms a circuit.

The basic processes of battery corrosion are both independent of each other and closely interconnected. If one process is hindered, the other two processes will also be affected. Therefore, to prevent the electrochemical corrosion of metals, it is necessary to reduce the corrosion rate from these three perspectives.

② Electrode Potential

Through the above example, it can be observed that the metal zinc plate underwent corrosion, while the metal copper plate did not undergo corrosion. Different metals exhibit varying degrees of susceptibility to corrosion. To explain this phenomenon, we need to provide a brief introduction to the theory of electrode potential.

Double Layer Theory: When a metal is immersed in an electrolyte solution, on one hand, there is a tendency for metal ions to enter the solution while leaving behind electrons. On the other hand, the metal ions in the solution are attracted by the electrons on the metal surface and have a tendency to deposit. These two opposing processes gradually reach a dynamic equilibrium, creating a potential difference between the metal surface and the electrolyte solution, known as electrode potential. The electrode potential reflects the ability of the metal to gain or lose electrons, and its magnitude depends on the characteristics of the electrode itself and is influenced by the electrolyte solution. Zinc has a lower standard electrode potential compared to

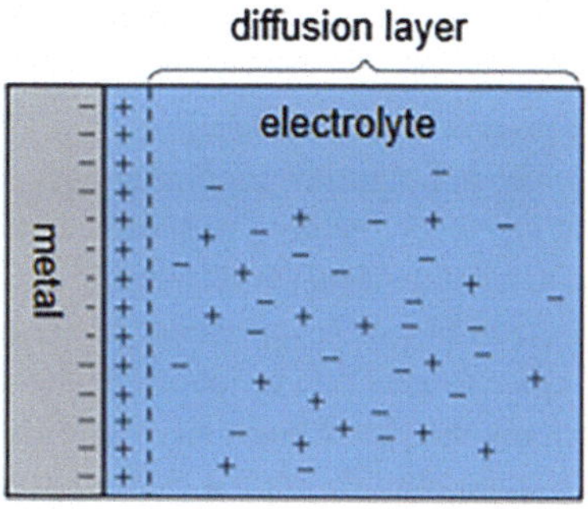

Fig. 3.17 Schematic diagram of the double-electric layer theory

copper, so zinc is generally used as the anode where it loses electrons and undergoes oxidation. The schematic diagram of the double layer theory is shown in Fig. 3.17.

The process of electrode gaining or losing electrons is not necessarily a reversible reaction. For example, when Fe corrodes in ($FeCl_3$) solution, the electrode gaining or losing electrons is not a reversible reaction, and it is called an irreversible electrode. The electrode potential of a reversible electrode is called an equilibrium electrode potential, while the electrode potential of an irreversible electrode is called a non-equilibrium electrode potential.

In a certain electrolyte solution, the potential of a metal or alloy electrode is mostly non-equilibrium. At this time, the directionality of the reaction should be determined based on the corrosion metal electrode ranking (electromotive series) of the metal in the corresponding electrolyte solution. In addition, the electrode potential of the metal is closely related to the pH value of the solution. The graph that describes the relationship between the electrode potential and the pH value is called a Pourbaix diagram.

③ Electrochemical Corrosion Tendency of Metals

According to the principles of thermodynamics, in a constant temperature and pressure system, the corrosion process of metals in an open system can be evaluated using the Gibbs free energy criterion. The relationship between the free energy of the reaction and the electrode potential of the metal can be expressed as follows [42]:

$$G = -nF(E_c - E_a) \tag{3.4}$$

Therefore, the thermodynamic conditions for metal corrosion can also be described as follows: The anodic electrode potential E_a of the metal is lower than the cathodic electrode potential E_c.

④ Polarization

The equilibrium electrode potential is measured when the circuit is open. When a corroding galvanic cell generates current, the electrode potential deviates from the equilibrium electrode potential. Generally, the cathode potential decreases while the anode potential increases, reducing the potential difference of the corroding galvanic cell and thus decreasing the corrosion rate. This phenomenon is referred to as polarization of the cell.

The polarization phenomenon is related to the various steps of electrode reactions. An electrode reaction generally consists of several consecutive unit steps: The mass transfer step of reactants passing from the bulk solution to the interfacial region; The electron transfer step where the reactants gain (or lose) electrons on the electrode surface to generate products; The process of mass transfer from the reaction product leaving the phase boundary region and dispersing into the solution or the process of gas generation is known as liquid-phase mass transfer. The lower limit of corrosion rate is mainly determined by the step with the greatest resistance of these steps.

The causes of polarization can be divided into three types: concentration polarization, electrochemical polarization, and resistance polarization.

- Concentration Polarization: The ions generated by electrode reactions need to be transported into the solution through convection, diffusion, and electromigration. Due to the slower transport rate compared to the generation rate of ions, the ion concentration near the electrode is inconsistent with the average concentration in the solution, resulting in a change in electrode potential.
- Electrochemical Polarization: In electrode reactions, the transformation of metal into metal ions requires a certain activation energy, which results in the transfer of metal ion charges being slower than the transfer of electrons. The accumulation of charges at the anode leads to a change in the electrode potential.
- Electrode Polarization: During the process of current passing through an electrolyte solution and electrode surface, electrode polarization occurs due to the presence of a certain type of membrane, resulting in an Ohmic potential drop at the electrode.

From the above polarization phenomenon, it can be seen that the electrode potential changes with the variation of corrosion current intensity. The graph showing the variation of anode and cathode potential with current intensity is called the Evans corrosion polarization curve. In a corrosion system with low resistance, corrosion with more drastic changes in anode potential compared to cathode potential is referred to as cathodic-controlled corrosion process, while corrosion with more drastic changes in cathode potential compared to anode potential is referred to as anodic-controlled corrosion process. If the polarization rates of both are similar, it is referred to as mixed anode and cathode control corrosion process. If resistance plays a dominant role, it is referred to as Ohmic resistance-controlled corrosion process.

⑤ Hydrogen Evolution Corrosion and Oxygen Absorption Corrosion

In theory, any substance capable of absorbing electrons generated by the anode can serve as the cathode of a corroded primary battery. Since the process of etching the cathode of the galvanic cell to absorb electrons is to reduce the cathode polarization, the oxidizing agent at the cathode can also be called a depolarizer. The two most common oxidizing substances are hydrogen ions (H^+) and dissolved oxygen (O_2).

The corrosion of metals using hydrogen ions (H^+) as depolarizers is known as hydrogen evolution corrosion. The necessary condition for hydrogen evolution corrosion is that the metal potential is lower than the hydrogen electrode. It is not necessarily related to the pH value of the corrosive medium. Metals such as iron and zinc, with lower potentials, can undergo hydrogen evolution corrosion in non-oxidizing acids without oxygen. Magnesium and its alloys, having even lower potentials, can undergo hydrogen evolution corrosion in alkaline solutions. Although metals like titanium and chromium are more active, they generally do not undergo hydrogen evolution corrosion due to their passivation.

The corrosion of metals using oxygen as a depolarizer is called oxygen absorption corrosion. Since the ionization potential of oxygen is 1.227V higher than that of hydrogen under the same solution and identical conditions, oxygen absorption corrosion takes precedence in acidic, neutral, or alkaline solutions. As the pH value of seawater generally ranges from 8 to 8.2, most metals used in marine environments undergo oxygen absorption corrosion. Magnesium, for example, undergoes both hydrogen evolution corrosion and oxygen absorption corrosion in seawater.

⑥ Passivation of Metals

The phenomenon of significantly enhancing the chemical stability of a metal or alloy under the influence of certain factors is called passivation of the metal. For example, the chemical properties of aluminum are too reactive and react with oxygen in the air to form a dense oxide film that prevents further oxidation.

Passivation of metals can be achieved by using chemical oxidants (such as concentrated sulfuric acid) or applying a direct current to the anode. The former is called chemical passivation, while the latter is called electrochemical passivation.

The corrosion system that can passivate automatically without relying on external anodic polarization current is called self-passivating. For example, stainless steel can be made by adding the easily passivated metal chromium to steel.

Because in the Cl^- rich seawater corrosion environment, Cl^- is easy to cause local damage to the passivation film, making it difficult for metals to form a passivation film to slow down corrosion. Therefore, the corrosion rate of offshore wind power metal equipment is much higher than that of onshore wind power metal equipment.

(3) Type

- Generalized corrosion: Corrosion that occurs uniformly across the entire surface of a metal, without distinguishable anodic and cathodic areas. The locations of corrosion may vary, such as the corrosion that occurs when a zinc metal sheet is immersed in dilute sulfuric acid. Although this type of corrosion leads to thinning of the metal, it does not necessarily imply uniform corrosion.
- Local corrosion: Compared to general corrosion, it is characterized by corrosion limited to or concentrated in a specific part of the metal. There are significant differences in corrosion between different parts, including galvanic corrosion, pitting corrosion, crevice corrosion, intergranular corrosion, exfoliation corrosion,

selective corrosion, filamentous corrosion, etc. Compared with comprehensive corrosion, local corrosion is limited to or concentrated in a specific part of the metal, and there are obvious differences among different parts of the corrosion. Although general corrosion can cause a large amount of metal loss, its harm is far less than that of local corrosion. This is because the rate of general corrosion is easy to measure, it is easy to be detected, and corrosion allowances can be considered in engineering design to prevent premature corrosion. It is difficult to predict and prevent local corrosion, often causing sudden failure of metal equipment without warning, leading to major engineering accidents or personal injuries.
- Corrosion caused by other factors: corrosion under stress, biological corrosion, ultraviolet radiation, etc.

2. **Corrosion of Reinforced Concrete**

The foundations of offshore wind turbines are typically made of steel or reinforced concrete materials. Here is a brief introduction to the mechanism of corrosion in reinforced concrete. In the marine environment, the main factors contributing to the corrosion and deterioration of reinforced concrete are chloride ingress, carbonation, sulfate attack, freeze–thaw damage, and the alkali-aggregate reaction in the concrete itself.

According to extensive research, in marine environments, chloride salts are the main cause of steel reinforcement corrosion in concrete.

(1) Effects of Chloride Salts

The corrosion of reinforced concrete can be divided into three main stages: corrosion initiation stage, corrosion propagation stage, and corrosion failure stage.

- Corrosion initiation stage: The high alkalinity of concrete forms a dense passivation film on the surface of the steel reinforcement. When chloride ions in the environment penetrate through the concrete and reach the steel surface, the passivation film can be disrupted, leading to corrosion. The time from the start of exposure to the initiation of corrosion is known as the corrosion initiation stage.
- Corrosion propagation stage: After the steel bar is corroded, it gradually loses the original material properties, which makes the bearing structure of the steel bar and concrete change, resulting in cracking of the concrete. This stage is referred to as the corrosion propagation stage.
- Corrosion failure stage: Cracks in the concrete structure increase the contact between steel reinforcement and the surrounding environment, accelerating the corrosion rate of the steel. Ultimately, this leads to structural failure. This stage is known as the corrosion failure stage.

(2) Carbonation Process

Carbonation of concrete refers to the reaction between the hydration products in cement paste and CO_2 in the environment, resulting in the formation of calcium carbonate or other substances. The deepening of carbonation degree can cause the surface of concrete to become soft and fragile. Under the impact of high-speed seawater waves, the surface concrete is prone to spalling, leading to a decrease in the effective thickness of the protective layer. For reinforced concrete structures, the dissolution of CO_2 in water will lower the pH value and promote the depassivation of reinforcing steel.

Carbonation is influenced by various factors such as environmental conditions and concrete manufacturing processes.

(3) Sulfate Action

Sulfate can promote metal corrosion in three aspects: First, it is similar to CO_2 to reduce the pH value of seawater and promote the corrosion of steel bars; The second is gypsum reaction with concrete structure; Third, sulfate can directly promote electrochemical corrosion of rebar.

(4) Other Factors

Concrete is influenced by factors such as the curing period, type of cement used, quality of aggregates, and admixtures.

3.4.2 Corrosion Environment and Corrosion Rate

1. **Characteristics of Seawater**

Seawater is a highly conductive and corrosive electrolyte that can dissolve oxygen. It primarily contains cations such as sodium ions, magnesium ions, calcium ions, and potassium ions, as well as anions such as chloride ions, sulfate ions, bicarbonate ions, bromide ions, and fluoride ions. Although the concentrations of many other ions and compounds in seawater are relatively low, their corrosive effects are significant.

Seawater also contains dissolved gases such as oxygen, nitrogen, and carbon dioxide. Oxygen-saturated surface seawater accelerates the corrosion rate of carbon steel, low alloy steel and other metal structures in seawater, and has a high corrosiveness. The periodic change of seawater temperature is also a factor affecting the corrosion of metals in seawater. Under normal circumstances, the corrosion rate of metals increases with the increase of temperature. In addition, a large number of Marine organisms and bacteria in the ocean will attach to the wind turbine, affecting its corrosion behavior.

2. **Corrosion Regions and Rates in Marine Environments**

The typical Marine environment can be divided into five typical corrosion areas from top to bottom: the marine atmosphere region, splash zone, tidal zone, submergence

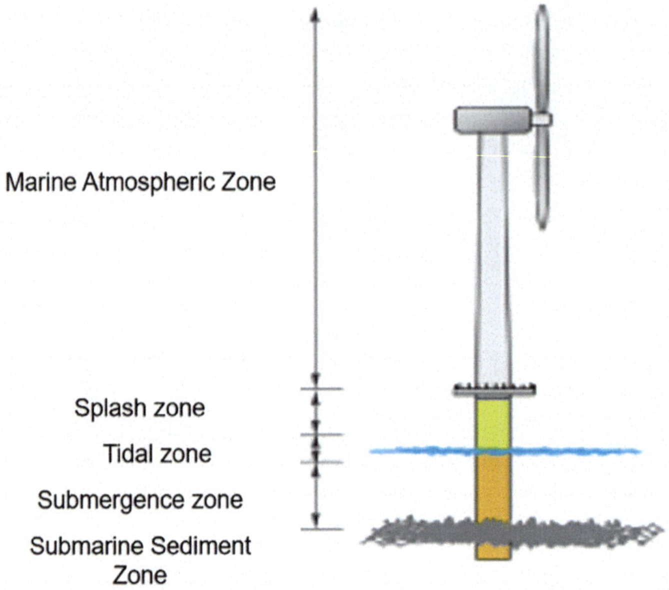

Fig. 3.18 Division of marine environment corrosion area

zone, submarine sediment zone. Offshore wind turbines (single pile foundation as an example) involve all of the above corrosion environment areas. The division of corrosion regions in marine environments is shown in Fig. 3.18.

(1) Marine Atmospheric Zone

The marine atmospheric zone refers to the area where there is no direct contact with seawater and is not reached by seawater splashes. It mainly corresponds to the area where the towers, nacelles, and blades of offshore wind turbines are located. Due to the high humidity and salinity of the ocean atmosphere, it is easy to form electrolyte solution films on the metal surface. The carbon atoms in the steel can be used as the anode, which is conducive to the formation of galvanic cells, promote electrochemical corrosion, and cause corrosion damage to the steel structure. At the same time, sodium chloride will form sodium chloride salt spray in the air with the evaporation of sea water, and can re-form sodium oxide solution after water. Salt spray corrosion will not only increase the corrosion rate, but also reduce the performance of wind turbines, increase the failure rate, and lead to more serious safety accidents [43].

This part of the corrosion is often affected by a variety of factors, in addition to the salt and moisture described above, but also including temperature, sunlight exposure, corrosion products and polluting gases.

(2) Splash Zone

The splash zone refers to the area where seawater can splash, which is above the average high tide line of the sea. The corrosion rate in the splash zone is much higher than in the marine atmospheric zone. On the one hand, the splash zone has a high concentration of sea salt particles, long duration of seawater immersion, frequent wet-dry cycles, and sufficient oxygen supply, promoting corrosion of steel structures through oxygen depolarization reactions. On the other hand, the scouring action of seawater can continuously remove the protective coatings and surface rust layers on the steel structure, exposing the steel surface to the corrosive environment. Based on the above two points, the corrosion rate of the splash zone is the largest in all areas, and the corrosion margin is generally selected for corrosion protection.

(3) Tidal Zone

The tidal zone refers to the area between the average high tide line and the average low tide line, which is exposed to air during low tide and submerged during high tide. In the tidal range of the sea water, the metal periodically comes into contact with the sea water and the air, and the oxygen supply is sufficient. The corrosion rate is low because it is used as the cathode in the macroscopic cell composed with the seawater immersion zone. The corrosion rate of metals in the tidal zone is usually affected by Marine organisms. In addition, in the tidal zone, the wind turbine may also be hit by floating ice and operation and maintenance ships.

(4) Submergence Zone

The submergence zone refers to the area below the average low tide line that is permanently submerged in seawater. In the submergence zone, steel structures such as the lower part of the wind tower platform are fully immersed in seawater. Since the pH value of seawater is close to neutral, the corrosion reaction of steel structures in seawater is generally oxygen absorption corrosion. In the submergence zone, dissolved oxygen plays a dominant role in the corrosion of steel structures.

In the upper part of the submergence zone, the corrosion rate is second only to the splash zone. The reason is that the upper part of the tidal zone and the submergence zone of the sea water together constitute the oxygen concentration battery. The oxygen content in the submergence zone is high, which is protected as the cathode of the oxygen concentration difference macroscopic battery. The rate of anodic corrosion is accelerated in the submergence zone with relatively low oxygen content, so the corrosion rate in this area is second only to the splash zone.

(5) Submarine Sediment Zone

The submarine sediment zone refers to the area within the submerged zone of the sea that is covered by sea mud. The corrosion characteristics of seafloor sediments vary with the depth of the sea and water. The surface layer of the submarine sediment zone is actually saturated seawater soil, which is fully mixed with seawater, with high salt content, low resistivity and low oxygen content, which is conducive to the reproduction of anaerobic bacteria. The by-products produced by bacterial activity may accelerate the corrosion of the material. In general, with the increase of depth, its corrosion ability also gradually decreases.

3. **Environmental Corrosion Levels**

Depending on the different regions where offshore wind turbines are located in the marine corrosion environment, the environmental corrosion levels also vary.

According to the classification standard of ISO 12944-2-2007 "Paints and varnishes—Corrosion protection of steel structures by protective paint systems—Part 2: Classification of environments", the environmental corrosion levels are classified from low to high corrosion as C1, C2, C3, C4, C5-I (Industrial), and C5-M (Marine). The environmental corrosion levels for water and soil are divided into three categories: Im1, Im2, and Im3, representing freshwater environments, seawater or brackish water environments, and soil environments, respectively.

In the marine atmospheric zone, the outer surface of wind turbines is directly exposed to the marine corrosion environment, and the environmental corrosion level can be classified as C5-M, while the inner surface, which does not directly contact the marine corrosion environment, can be classified as C4 or C5. If environmental control is implemented within the nacelle, it can also be classified as C3 or below.

In the splash zone and tidal zone, the environmental corrosion level is generally classified as Im2. In the submergence zone, the environmental corrosion level is classified as Im2. Below a certain depth in the submarine sediment zone, the environmental corrosion level of the outer surface of steel piles can be classified as Im3.

For the classification standards, environmental conditions, corrosion characteristics, and corrosion levels of different marine zones, please refer to Table 3.8.

3.4.3 Corrosion Protection Design Forms

The service life of offshore wind turbines is generally required to be no less than 20 years, with a foundation life of no less than 25 years. Based on the corrosion mechanisms discussed earlier, corrosion protection methods can generally be divided into isolation protection, electrochemical protection, and intrinsic protection. In wind turbines, corrosion rates can also be reduced by strengthening sealing.

The forms of corrosion protection design for offshore wind turbines mainly include anti-corrosion coatings (such as anti-corrosion paints, metal thermal spraying, and plating), cathodic protection, encapsulation isolation, structural design, corrosion allowance, environmental control (heating and dehumidification, and enhanced sealing), material selection, and rust prevention oils. Among these, material selection, structural design, and anti-corrosion coatings are widely applied in

Table 3.8 The classification standards, environmental conditions, corrosion characteristics, and corrosion levels of different marine zones [44]

Marine zones	Classification standards	Environmental conditions	Corrosion characteristics	Corrosion levels
Marine atmospheric zone	The area where there is no direct contact with seawater and is not reached by seawater splashes	Small sea salt particles are brought by the wind, and the influencing factors are height, wind speed, rainfall, temperature, radiation, etc.	Sea salt particles increase the corrosion, depending on the distance from the shore	External: C5-M; Internal: C5-M/C4
Splash zone	The area where seawater can splash, which is above the average high tide line of the sea	Wet, fully inflated surface, no marine contamination	Sea water splashing, dry and wet alternately, severe corrosion	Im2
Tidal zone	The area between the average high tide line and the average low tide line	Cycle immersion, adequate oxygen supply	The battery is protected by the oxygen concentration	Im2
Submergence zone	The area below the average low tide line that is permanently submerged in seawater	The shallow sea is saturated with oxygen, and the influencing factors include oxygen content, flow rate, water temperature, Marine life, bacteria, and so on	Corrosion changes with temperature and seawater depth, and the cathode region forms gray matter, greatly affecting biological factors	Im2
Submarine sediment zone	The area within the submerged zone of the sea that is covered by sea mud	Often have bacteria, such as sulfate-reducing bacteria	Mud is usually corrosive, causing microbial corrosion	Im2

the corrosion protection of foundations, towers, and nacelles. Corrosion allowance is mainly used for the corrosion protection of splash zone and tidal zone of seawater. Metal thermal spraying is mainly used for corrosion protection of tower bodies, flange faces, and engine compartments in marine atmospheric zone C5-M. Cathodic protection is mainly used for basic anti-corrosion in Im2/Im3. Coating and isolation are mainly used for basic anti-corrosion in splash and tidal zones of seawater with Im2. Heating and dehumidification are mainly used for corrosion protection of tower structures and internal electrical equipment in engine compartments.

1. **Anti-corrosion Coating**

Anti-corrosion coating mainly includes anti-corrosion coatings (mainly anti-corrosion paint) and metal coatings (including metal thermal spraying and plating).

The anticorrosion mechanism of anticorrosive coatings: It forms a protective layer on the metal surface to prevent direct contact between the metal and corrosive media, thereby reducing the corrosion rate. Due to the high maintenance cost and strong environmental corrosion of offshore wind turbines, traditional anticorrosion measures are not sufficient to meet the anticorrosion requirements. Therefore, heavy-duty anticorrosive coatings are commonly used. Heavy-duty anticorrosive coatings generally have the characteristics of thick film coating, high performance, and high surface treatment requirements for the substrate.

Metal thermal spraying belongs to the coating method, which involves heating metal spraying materials to a molten state and using the power of airflow and other forces to impact and deposit them on the surface of the substrate that needs protection, forming metal coatings with required properties. Common metal spray materials include zinc, stainless steel, etc. Among them, stainless steel coatings have the characteristics of wear resistance and long protection cycle. Zinc coatings don't only have covering and corrosion resistance functions, but more importantly, they also have cathodic protection functions [45].

The coating method belongs to the isolation corrosion protection and is mainly used in marine atmospheric zones, splash zones, and tidal zones. This method is commonly used for smaller auxiliary components or connection components in offshore equipment. The commonly used anti-corrosion coating methods include hot-dip galvanizing and electroplating. Hot-dip galvanizing involves immersing surface-treated structural parts into a bath of molten coating salt at high temperature, utilizing displacement reaction to form a metal protective layer with the designed thickness on the surface of the structural parts. Electroplating, on the other hand, uses an applied electric current for electrolytic displacement.

2. **Cathodic Protection**

Cathodic protection is a method of electrochemical corrosion protection. It can be divided into impressed current cathodic protection and sacrificial anode cathodic protection. Essentially, both methods reduce the corrosion rate by changing the electrode potential.

Impressed current cathodic protection is mainly forced to form a potential difference by an external power source. The material that is difficult to corrode is used as the anode, and the structure that is easy to corrode is used as the cathode to prevent corrosion. Sacrificial anode cathodic protection primarily utilizes materials with higher reactivity, such as zinc and aluminum, as the anode. These materials are connected to the protected structure through welding or other methods to form a protective potential difference and prevent corrosion.

3. **Cladding Isolation**

Cladding isolation mainly uses other materials to isolate the protected structure from the environment to achieve the purpose of slowing down corrosion. For example, traditional concrete cladding; corrosion-resistant Monel local cladding; the new layer of mineral grease coating (PTC) proposed by Academician Hou Baorong and others in the ocean steel structure spray splash zone.

4. **Structural Design**

Appropriate structural design can reduce corrosion rates, such as by designing wind turbines to better withstand salt spray and seawater, minimizing contact with corrosive media. Structural design is commonly applied to offshore wind turbine systems, offshore substation foundations, and tower sections for corrosion protection.

5. **Corrosion Allowance**

Corrosion margin refers to the artificial thickening of corroded parts so that they do not suffer structural failure during their design life. Its value is determined by the corrosion rate of the medium to the material and the design life of the parts. It is generally used only for parts of lesser significance, structures with short design lives, and areas where regular inspection and repair work is performed.

For example, in ISO 19902-2007 "Petroleum and natural gas industries—Fixed steel offshore structures," if the specified splash zone corrosion rate is 0.3 mm/year and the design life is 20 years, the total corrosion allowance for the splash zone would be 6 mm.

6. **Environmental Control**

Environmental control mainly includes heating dehumidification and strengthening seals to prevent corrosion by reducing the corrosiveness of corrosive media.

- Heating and dehumidification design: When the metal surface is wet, electrochemical corrosion is easy to occur, and the corrosion rate is faster. But in a dry environment, the rate of pure chemical corrosion is slower. Therefore, the corrosion rate can be reduced by sealing and installing air conditioning inside the wind turbine to keep the internal air dry.
- Enhanced sealing design: For main engine covers such as the engine cowling and deflectors, as well as rotating components like the main bearings, gearbox, and generator, corresponding sealing measures should be implemented to reduce salt spray and water vapor into the wind turbine interior.

7. **Material Selection**

Improving the corrosion resistance of the material itself is one of the important means to avoid corrosion failure. However, the replacement of materials not only needs to consider the difficulty of design and manufacturing, but also needs to meet technical and economic requirements. For example, adding a certain amount of manganese, chromium, phosphorus, alum and other elements in the smelting of ordinary carbon steel to make the corresponding alloy can improve the corrosion resistance, but its mechanical properties and material costs will also change.

3.5 Lightning Protection Design

The individual capacity of offshore wind turbines is often larger than that of onshore wind turbines. Increasing the individual capacity of wind turbines leads to longer blades and a greater overall height of the turbines. Due to the frequent thunderstorms at sea and the absence of obstructions, offshore wind turbines are more vulnerable to lightning strikes. Furthermore, maintenance is more challenging and expensive because of the harsh environmental conditions and the turbines' susceptibility to corrosion. Consequently, the construction and maintenance costs of offshore wind turbines are significantly higher than those of onshore wind turbines, which imposes stricter reliability requirements. Lightning protection and grounding systems play a crucial role in the reliability of wind turbines. Lightning strikes can cause substantial damage to wind turbines, and inadequately designed grounding systems can severely compromise the operational stability of the turbines [46]. Given the unique environmental conditions in which offshore wind turbines operate, their reliability demands are much greater than those of onshore wind turbines.

3.5.1 Formation and Hazards of Lightning

Lightning is an atmospheric discharge phenomenon. Particles such as dust and ice crystals in the air become charged through vigorous friction during motion. Cloud masses intersect magnetic field lines, leading to the accumulation of positive charges in certain parts of the cloud and negative charges in other parts. When the air between centers of opposite charges is punctured by the strong electric field during their movement, a discharge occurs. Such discharges can take place between clouds or between clouds and the Earth. The phenomenon of wind turbines being struck by lightning essentially involves a discharge between charged thunderclouds and the Earth. Among all forms of lightning discharges, both the positive polarity discharge from thunderclouds to the Earth and the negative polarity discharge from the Earth to thunderclouds carry substantial electrical currents and energy. Lightning strikes can cause various failures in wind turbines. The rate of damage to key components of wind turbines, ranked from highest to lowest, is as follows: control system, blades, and generator. The maintenance costs resulting from lightning strikes follow the order from highest to lowest as blades, generator, control system, and communication system. Among these, the most severe hazard posed by lightning is the ignition of wind turbines, which can lead to considerable damage.

Table 3.9 Average wind turbine downtime and average power loss due to faults. (Including the number of lightning events)

Faults	Number of lightning events	Average Wind Turbine Downtime (h)	Average power loss (kW/h)
All faults	10,192	91	2249
Lightning strike faults	461	110	3200
Rate of Deviation from the Average Faults	/	+ 20.8%	+ 40.3%

During regular operation of wind turbines, lightning strikes not only result in equipment damage but also lead to significant downtime. According to the international standard IEC 61400-24, statistics indicate that downtime caused by lightning strikes (mainly including the repair period of the electrical system, the time for component procurement and transportation, etc.) is the longest among various types of wind turbine failures. Table 3.9 presents the average wind turbine downtime and average power loss caused by different faults. The statistics reveal that lightning faults, although they occur less frequently in the total number of faults in wind turbines, cause more downtime than all other faults combined. The periods of downtime can result in a considerable loss of power generation and thus the corresponding economic losses.

Lightning overvoltage can cause four types of damage to wind turbines:

- Direct lightning damage caused by direct lightning flashes on equipment.
- Ground potential backlash damage to wind turbines caused by transient ground potential raising at the contact resistance when lightning currents flow into the grounding body.
- Lightning power intrusion wave damage caused by lightning overvoltage intruding into the equipment along the power and signal lines connected to the equipment.
- Due to improper installation of the device, the sensitivity of the component has been adversely affected by the impact of lightning's electric or magnetic fields.

3.5.2 *Mechanisms of Lightning Damage*

The mechanisms of lightning-induced damage to wind turbines are closely related to the current waveform and lightning parameters of the lightning discharge. Lightning parameters include peak current, transferred charge, and current steepness.

1. **Peak Current**

When lightning current passes through the struck object, it results in an elevation of temperature within the object. In the case of wind turbine blades, their damage is often closely related to this thermal effect. Fundamentally, this thermal effect is associated with the energy contained in the lightning discharge, wherein the peak current plays a critical role. As the lightning current flows through the struck object, such as the conductor of a blade, it can generate substantial electromagnetic forces, which may cause bending or even fracture of the blade. Additionally, the lightning current channel may give rise to an electric arc. The overvoltage expansion produced by the arc is related to the integration of the lightning current waveform. The sudden increase in high temperature during arc burning can result in severe damage to the struck object. This constitutes a primary cause of blade damage.

2. **Transferred Charge**

When an object is struck by lightning, most of the charge transfer occurs during the process of relatively long duration and relatively low amplitude lightning currents. Prolonged lightning currents cause localized surface melting of the struck object, leading to the formation of erosion spots. Along the path of the lightning current, once an electric arc is formed, erosion spots emerge at the location of the arc occurrence. If the lightning current is sufficiently large, it may even result in metal melting. This poses a potential threat to the safety of wind turbine bearings since the occurrence of electric arcs is highly probable at the bearing contact surfaces, possibly leading to bearing welding. Even in the absence of bearing welding, the erosion spots on the bearings accelerate wear and shorten their lifespan.

3. **Current Steepness**

Lightning strikes on wind turbines often cause damage to control systems or electronic components, primarily due to the existence of induced over voltages. The induced overvoltage is closely related to the steepness of the lightning current waveform. The steeper the lightning current waveform, the higher the induced overvoltage.

3.5.3 Lightning Protection Design Principles and Definitions

The design of lightning protection systems should be integrated with the structural design of offshore wind turbines and other facilities to achieve an optimized solution that considers technical, economic, and safety factors. In particular, relevant metallic components can be utilized as integral parts of the lightning protection system in the design of offshore wind turbines and structures [47].

1. **Principles of Lightning Protection Design**

The risk of damage to offshore wind turbines from direct lightning flashes is reduced by selecting the appropriate level of protection required for the maximum permissible conditions.

In accordance with the electromagnetic requirements, offshore wind turbines should be divided into several lightning protection zones, each necessitating the implementation of measures tailored to its specific characteristics [47].

2. **Definitions**

- Lightning Protection System: A system designed to mitigate physical damage and personal injuries caused by lightning strikes on or in the vicinity of buildings or structures.
- Lightning Electromagnetic impulse: The electromagnetic effects generated by lightning current coupling through resistance, inductance, and capacitance. It encompasses both lightning surges and radiated electromagnetic fields.
- Lightning Protection Zone: An area delineated based on the electromagnetic intensity of lightning strikes. The boundary does not necessarily require physical barriers such as walls, floors, or ceilings. Once established, the electromagnetic pulse intensity in the protected zone should be matched with the internal system's tolerance to withstand the impacts.
- Lightning Level: A set of probabilistic parameters of lightning currents used for designing lightning protection measures. Natural lightning currents that exceed the maximum or fall below the minimum values in this set are low-probability events, which can reflect the effectiveness of the lightning protection system.
- Lightning Equipotential Bonding: This lightning protection measure aims to minimize potential differences caused by lightning currents. It involves directly connecting separate metallic components with short, straight conductors or using surge protectors to connect them to the lightning protection system.
- Lightning Surge: A transient wave characterized by overvoltage and/or overcurrent resulting from lightning electromagnetic pulses. It can originate from (localized) lightning currents, induction effects in facility internal loops, residual threats from downstream surge protectors, or other causes such as operation of switches or fuse trips.
- Surge Protective Device: A device used to limit transient overvoltage and dissipate surge currents. It contains at least one nonlinear component.
- Lightning Receptor: Components of a lightning protection system designed to intercept lightning strikes. They include lightning rods, lightning tapes, lightning conductors, lightning meshes, metallic surfaces, and metal structures that intercept lightning strikes.
- Lightning Receptor Tapes: A type of air terminal that is horizontally or diagonally installed (based on the inclination of the building surface) and connected to down conductors at two or more points (e.g., the beginning and end of the tape). If suspended, it is referred to as a lightning conductor, while a mesh-like installation is called a lightning mesh.
- Down Conductor System: Conductors used to conduct lightning currents from air terminals to grounding devices.
- Busbars: Also known as grounding copper bars, they establish equipotential connections between grounding conductors and various metal components within

the system. The voltage difference within the equipotential system is almost zero, providing safety for personnel and equipment during overcurrent or faults. Busbars are typically installed at the front end of the grounding network from the equipment room, where all equipment grounding terminals converge. They are primarily used for voltage equalization in internal lightning protection or electrical systems.

- Earth Electrodes: Conductors buried in the soil or placed on concrete foundations to serve as dispersal conductors.
- Earth-termination System: The combination of grounding electrodes and grounding conductors used to conduct lightning currents and disperse them into the ground.
- Foundation Grounding Electrode: Additional conductors such as reinforced steel bars in the foundation concrete or embedded in the building's concrete, serving as grounding electrodes.
- Low-Voltage Electrical System: Includes the main electrical power system (wind turbine generator, converter, and related components) used for generating and transmitting wind power energy, as well as the low-voltage power supply system used for measurements.
- Lightning Arrester: An electrical component designed to protect equipment from high transient overvoltage while limiting both the duration and amplitude of these surges. Typically, it is connected between the power grid wire and the ground wire; it may also be connected to the winding of electrical equipment or between wires. This component is commonly referred to as an overvoltage protector or overvoltage limiter.
- Protective Grounding: Grounding performed to prevent the electrically charged metal enclosures of electrical devices, the framework of distribution equipment, and line towers from posing risks to personal and equipment safety. It involves reliably connecting the electrically conductive metal parts (insulated from the charged parts) that are normally uncharged but may become charged due to insulation damage or other circumstances to grounding electrodes using conductors. It is a protective wiring method.
- Functional Grounding: Grounding employed to ensure the proper operation of equipment or systems and to facilitate their reliable and correct functioning [48].

3. **Determination of Protection Level**

Although offshore wind turbines are classified as general industrial constructions, their location at sea level, height exceeding 100 m, and the absence of nearby tall buildings make them particularly susceptible to lightning strikes in areas prone to frequent thunderstorms. Therefore, it is crucial to accurately assess the lightning protection requirements for offshore wind turbines during the design phase.

The annual estimated number of lightning strikes on wind turbines is given by the equation:

$$N = k \times N_g \times A_e \tag{3.5}$$

where:

N represents the annual estimated number of lightning strikes on wind turbines (strikes/a);

k is the correction factor (for offshore wind turbines, it is taken as 2);

N_g represents the annual average density of lightning strikes to the earth in the wind farm area (strikes/($m^2 \cdot a$));

A_e represents the equivalent area (in m^2) that intercepts the same number of lightning strikes as the wind turbine.

The shape of a wind turbine is irregular, with the highest height H where the blades are located, the longest length L for the nacelle, and the widest diameter W for the rotor. Therefore, when calculating, the wind turbine should be approximated as a regular geometric shape with a height of H, a length of L, and a width of W. Since the height of the wind turbine is greater than 100 m and there are no obstructions nearby, the expanded width on each side should be calculated based on the height of the wind turbine.

According to the calculated annual estimated number of lightning strikes for the wind turbine, as specified in GB 50057-2010, the lightning protection level of the wind turbine can be determined. For instance, if the wind turbine experiences N = 0.25 strikes/a, it should be classified as having a level II lightning protection level. In the design specifications, the lightning current amplitude is set at 150 kA, with a wave front time (T1) of 10 μs, a half-value time (T2) of 350 ps, and a ball radius (R) of 45 m.

4. **Vulnerable Locations and Lightning Protection Design for Wind Turbines**

When thunderclouds discharge to the ground, lightning is more likely to strike the tips of a wind turbine's blades. However, it is also possible for lightning to strike the sides of the blades, the insulated components, or even the internal conductors. The discharge from the ground to the thundercloud initially forms at the highest points, which are typically found at the blade tips and other external protrusions, such as lightning rods on the nacelle, the front end of the nacelle, and the hub. If the blades are equipped with lightning protection at the tips, the upward-propagating lightning discharge will also concentrate at these points. Consequently, when a wind turbine is struck by lightning, the lightning strikes may be distributed across various locations.

According to analysis, the damage caused by lightning is mainly due to the powerful current generated by atmospheric discharge, which is related to parameters such as the maximum discharge current and discharge charge. Currently, there are several common measures taken for lightning protection in wind turbines.

(1) Lightning protection measures for vulnerable areas

Vulnerable areas are at high risk of lightning strikes, and the key to lightning protection in these areas is to ensure that the lightning current can be quickly conducted. The most commonly used measure is still the traditional Franklin lightning protection method. Additionally, the height of vulnerable areas can be utilized to distort the electric field of the thundercloud, protecting the target by locally replacing it as the lightning strike point. For wind turbines, the blades are located at the highest position and are in motion, which imposes higher requirements on the materials used for blade construction and the discharge channels for lightning current.

(2) Equipotential Bonding

All electrical conductors of offshore wind turbines must be connected to an equipotential system. Typically, the cables used for this equipotential connection are flat and are commonly referred to as metal connection strips. The electrical potentials at both ends of the metal bonding strap must be equal to effectively reduce electromagnetic interference. Additionally, a reliable electrical connection should be established between the rotor, nacelle, nacelle base, and tower. The resistance of each connection should be minimized and should generally not exceed 0.03 Ω.

(3) Installation of Surge Protection Devices

Electrical circuits and communication lines can experience external interference, such as lightning, which produces a peak current or voltage lasting more than 3×10^{-9} s. This phenomenon is referred to as a surge. A surge protection device can activate in a very short time and function as a shunt, thereby preventing damage to electronic equipment in the electrical circuit caused by the surge's impact. In lightning protection assessments, wind turbines are generally classified as B-level protection. For electronic equipment within B-level protection areas, it is required to install three-level surge protection devices: the first level being switch-based surge protection devices, and the second and third levels being voltage limiting surge protection devices. Various control cabinets and switch cabinets in wind turbines should be equipped with such devices.

(4) Isolation

In general, optoelectronic isolation can effectively reduce and suppress various interferences generated during the transmission of weak electrical signals, enabling reliable and stable operation of electronic equipment. For wind turbines, employing optical fiber connections between the processor in the nacelle and the controller on the ground can effectively mitigate interference.

3.5.4 Lightning Protection Measures

1. **Lightning Protection for Wind Turbine Blades and Nacelles**

When wind turbine blades are struck by lightning, common forms of damage include blade cracking, burning, and degradation of the composite material surface, as well as the destruction and melting of metal components. The temperature variation on the exterior of the blades causes the moist air trapped inside to condense, leading to the accumulation of water droplets at the blade tips due to the blades' rotation. When lightning strikes the blades, the water vapor and air inside rapidly expand, creating an instantaneous and tremendous pressure shock that can cause the entire blade to rupture. In severe cases, this pressure can be transmitted through the hub to unaffected blades, resulting in chain damage. Therefore, the key to protecting blades from lightning is to minimize the presence of water droplets at the blade tips. The standard practice is to install a drainage hole at the blade tip and place a lightning receptor at this drainage hole. Additionally, the quality of wind turbine blades is determined by the quality of their components. Consequently, before the blades leave the factory, it is essential to test the resistance between each lightning receptor and the metallic flange at the root of the blade. When the resistance reaches the milliohm level, it indicates a good quality connection of the lightning receptor at that time.

2. **Lightning Protection for Wind Turbine Bearings and Gearboxes**

Wind turbines commonly contain components such as yaw bearings, pitch bearings, and gearboxes with speed-increasing functionality. The motion and contact surfaces of bearings vary during operation, and when the current density passing through the bearing contact surface exceeds a certain threshold, electrical arcing occurs. In particular, when lightning strikes the wind turbine blades, a massive current flow through the bearings and gearboxes, leaving burn spots on the contact surfaces. This results in increased vibration and mechanical wear in the bearing and gear system, thereby reducing their operational lifespan. Therefore, in terms of lightning protection for wind turbine bearings and gearboxes, it is important to consider methods of reducing the current flowing through the bearing contact surfaces when the blades are struck by lightning. In practical engineering, following the guidelines of standards as IEC 61400-24, insulation layers are typically added to alter the path of lightning current, and slip rings parallel to the bearings are installed to handle a portion of the lightning current. Currently, most wind turbine slow-speed shafts and gearboxes are fixed to the nacelle base through brackets, with insulation layers between them. Additionally, the flexible coupling between the high-speed shaft of the gearbox and the generator's drive shaft also serves as an insulating element.

3. **Internal Lightning Protection for Offshore Wind Turbines**

The internal components of offshore wind turbines consist of power circuits and signal circuits, which are interconnected with electrical equipment. The primary objective of internal lightning protection in offshore wind turbines is to mitigate the effects of lightning currents on the circuits and electrical equipment. Lightning currents tend to follow the path of least resistance, and when they pass through power and signal circuits, as well as the connected electrical equipment, they can cause damage. Within the wind turbine, the passage of lightning currents generates alternating magnetic fields. Since the power and signal circuits run parallel to the path of the lightning current, they are subject to induced over voltages. The magnitude of these induced over voltages can reach several tens of kilovolts, thereby affecting and damaging the power and signal circuits. Additionally, lightning currents can create residual voltages on the grounding system, raising the ground potential and potentially impacting the stability of the wind turbine's control system.

(1) Prevention of direct injection of lightning currents

The direct injection of lightning currents into relevant circuits and equipment is a severe phenomenon that can cause significant damage. Therefore, preventing lightning currents from entering relevant circuits and electrical devices is the primary task of lightning protection. Achieving this task is relatively straightforward. It requires the scientific arrangement of lightning rods, down conductors, grounding wires, and busbars, and ensuring their reliable connections to ensure that lightning currents can flow to the down conductor through the shortest path. The lightning current first flows through the down conductor connected to the grounding body because this path is the shortest and has the lowest resistance. It then flows to the grounding body, thereby ensuring that the lightning current does not enter other circuits and electrical devices. For signal circuits of anemometers and wind vanes protected by lightning rods, the signal lines should be routed along metal components and shielded and insulated to prevent the injection of lightning currents.

(2) Suppression of induced over voltages

The induced over voltages generated by lightning strikes can damage the internal power and signal lines of wind turbines, as well as the electrical equipment connected to them. These damages can be potential, potentially leading to major faults during future operation, or they can be manifest, causing instability in the operation of the wind turbines. Therefore, suppressing induced over voltages is one of the most challenging aspects of lightning protection. The general methods for reducing induced over voltages are as follows.

- Power and signal lines should be kept as short as possible and located close to metal components. Low-voltage and high-voltage cables should maintain a certain safety distance.
- For long-distance control lines, fiber optic communication should be used whenever possible.
- Shielded cables should be employed for long-distance auxiliary power lines within the tower, running from top to bottom, with both ends grounded. All non-fiber optic signal lines should use shielded cables. When transmitting low-frequency signals, the ends of the shielding layer should be reliably grounded. When transmitting high-frequency signals, the ends of the shielding layer should be grounded at the closest point to the grounding body to ensure maximum stability of control signals and power supply.
- Sensitive control lines should be arranged in fixed cable trays at both ends.

- In the wind turbine controller, surge protectors of corresponding voltage levels should be used for communication lines of different voltage levels of power transformers to prevent overvoltage. However, for the 24V DC power supply and I/O modules in the controller, limiting surge protectors should be used.

(3) Suppression of lightning current-induced ground potential rise on the control stability of wind turbines

Grounding for wind turbine units can be classified into protective grounding and functional grounding, each serving specific purposes. It is challenging to completely separate all types of grounding in practical wind turbine installations. The purpose of lightning protection grounding is to direct the lightning current into the ground. This process significantly affects the ground potential and may elevate it, consequently impacting the functionality of other grounds. Therefore, the lightning protection grounding for wind turbine units should be implemented separately. If conditions allow, it is advisable to separate different grounds as much as possible to ensure that each ground is minimally affected by others. Please refer to Table 3.10 for types of grounding.

The grounding system of offshore wind turbines consists of a circular conductor that encircles the foundation. This conductor is positioned 1 m below the water surface or sediment, at a specified distance from the foundation. It utilizes a copper conductor with a cross-sectional area of 50 mm^2 or a larger diameter, ensuring it remains submerged beneath the water surface. Cathodic protection is implemented using sacrificial anodes, and multiple grounding terminals are connected to accommodate varying ground conditions.

Table 3.10 Grounding types for offshore wind turbines

Category	Grounding name	Purpose
Protective grounding	Protective grounding	Ensure personal safety
	Lightning protection grounding	Lightning current introduction ground
	Static grounding	Introduce static charges into the ground
	Corrosion protection grounding	Cathodic protection
Functional grounding	Operational grounding	Ensure proper system operation
	Logic grounding	Obtain reference points
	Signal grounding	Obtain baseline reference points
	Shielding grounding	Prevent electromagnetic interference

3.5.5 *Inspection and Maintenance of Lightning Protection Systems*

1. **Content of Lightning Protection System Inspection**

- Consistency of the lightning protection system with the design.
- Good condition of all components of the lightning protection system, with the capability to fulfill the intended functions as designed, and no corrosive effects.
- For any newly added offshore wind turbines or facilities, the newly added lightning protection system should be connected to the existing one or the existing lightning protection system should be extended to incorporate the new facility into the previously protected space [47].

2. **Procedure for Lightning Protection System Inspection**

- During the installation and construction period of offshore wind turbines and other facilities, the buried grounding elements should be checked.
- After the lightning protection system is installed, an inspection should be conducted according to the requirements stated in point 1 of this section.
- Regular repeated inspections should be performed as required in point 1 of this section, with the inspection intervals determined based on the characteristics of the protected space and the corrosion conditions.
- Additional inspections should be conducted in accordance with the requirements stated in point 1 of this section in case of any changes, repairs, or lightning strikes affecting the offshore wind turbines.

3. **Maintenance**

- Regular inspection of the lightning protection system is a fundamental requirement for reliable operation and any detected defects should be promptly repaired.
- The operation and maintenance section of the user manual for offshore wind turbines should include specific instructions for lightning protection system inspection and maintenance [47].

References

1. Wang ZQ. Establishment and implementation of an intensive integrated management system. China Petrochemical Press; 2010: 21–22.
2. Chen XH, Zhang XG, Li RF. Design and development of offshore wind turbines. Beijing: China Electric Power Press; 2018.

3. Fu PC, Liu JP, He KH, et al. Analysis of difficulties in "integrated design" of offshore wind power projects. Wind Energy. 2020;120(02):70–1.
4. Zhai ED, Zhang XG, Li RF. Integrated design of offshore wind turbine tower foundation. Southern Energy Construction. 2018;5(02):1–7.
5. Kaufer D, Cosack N, Böker C, et al. Integrated analysis of the dynamics of offshore wind turbines with arbitrary support structures. In: European wind energy conference, Marseille, France. 2009.
6. Gentils T, Wang L, Kolios A. Integrated structural optimization of offshore wind turbine support structures based on finite element analysis and genetic algorithm. Appl Energy. 2017;199:187–204.
7. Popko W, Vorpahl F, Jonkman J, et al. OC3 and OC4 projects—verification bench-mark exercises of state-of-art coupled simulation tools for offshore wind turbines. In: The 7thEuropean Seminar Offshore Wind and other Marine Renewable Energy in Mediterranean and European Seas (OWEMES), Rome, Italy, 2012, 499–503.
8. Zhao X, Fan Y, Xia J, et al. Reliability design of large offshore wind turbines. Explosion-proof Electric Mach. 2019;54(004):16–23.
9. Cui YL. Review of reliability design methods for mechanical products. Mach Des. 2009;26(12–13):22.
10. Hou Y. Introduction of reliability design (IV). Tech Supervision Pet Ind. 1999;7:34–6.
11. Spinato F, Tavner PJ, Bussel GV, et al. Reliability of wind turbine subassemblies. IET Renew Power Gener. 2009;3(4):387–401.
12. Xie SF. Reliability design. Beijing: Publishing House of Electronics Industry; 2015.
13. Wu JL, Li CF. Offshore wind power generation technology. Beijing: Chemical Industry Press; 2010.
14. Zhou T, Jin B, Li XP, et al. Reliability generalized margin design method for complex systems. In: Proceedings of the 4th China Aerospace Quality Forum, 2013.
15. Sun WW, Zhou H. Effective redundancy design. Electron Prod Reliab Environ Test. 2008;26(3):47–50.
16. Guo ZF, Guo B, Zhao K. Research on derating design of electronic components. Electron Technolo Softw Eng. 2016;1:257–8.
17. Chen YM, Yu DZ, Huang BX. Thermal design (II). Electron Technol. 1981;12:32–4.
18. Peng DY, Liu CH, Zhao SP, et al. Review of reliability design methods for electronic equipment. In: Proceedings of the 4th symposium on quality and reliability of electronic information systems, 2008.
19. Liu HJ, Zhao HM, Wang CY. Mechanical reliability design. Beijing: National Defense Industry Press; 2009.
20. Chen C. Effect mechanism of crosswind Wind Effect on safety margin of fan expansion foundation. Yichang: China Three Gorges University; 2013.
21. Li WR, Zhu JC. Discussion on locking technology of threaded fasteners [J]. Development and Innovation of Mechanical and Electrical Products. 2003;000(002):15–7.
22. Fang JY. Research on strength reliability optimization of wind turbine gearbox. Beijing: North China Electric Power University; 2013.
23. Liu D. Reliability analysis and optimization design of wind turbine gearbox. Chengdu: University of Electronic Science and Technology of China; 2020.
24. Wang XW. Mechanical system reliability basis. Beijing: China Machine Press; 2019.
25. Zeng SK, Zhao YD, Zhang JG. System reliability design analysis course. Beijing: Beijing University of Aeronautics and Astronautics Press; 2001.
26. Song N. Data-based generator failure prediction and Health Management of wind turbine. Shanghai: Shanghai Jiao Tong University; 2020.
27. Zhang T. Reliability, availability and maintenance design of offshore wind turbines. Beijing: China Machine Press; 2018.
28. Xue S. Research on reliability evaluation and improvement measures of DC capacitor in wind power converters. Chongqing: Chongqing University; 2017.

29. He GL, Tian JK, Chang DS. Conceptual design of typhoon resistance of offshore wind turbine. Plann Des. 2013;34(2):11–7.
30. Zhang F, Wu QH, Li JH. Damage and preventive measures of Zhejiang Power grid caused by Typhoon "Yunna." China Electric Power. 2005;38(5):39–42.
31. Liu HT. Research on typhoon resistance strengthening design of 1 MW wind turbine. Beijing: North China Electric Power University; 2015.
32. Chen JS, Zhang B. Analysis of typhoon resistance of offshore wind turbine [J]. Research Park, 72–75.
33. He GL, Li J. Wind dynamic reliability analysis of wind power tower system. Journal of Tongji University: Natural Science Edition. 2011;39(4):8.
34. He GL, Li J. Wind-induced dynamic response analysis of high tower system for wind power generation. J Electric Power Constr. 2011;32(10):1–9.
35. Zhang QH, Yang S. Research on single reinforced structure of engine room of megawatt wind turbine. Wind Energy Industry. 2014;11:3.
36. Xiang S. Research on vibration control technology of flexible tower of offshore wind turbine. Shenyang: Shenyang University of Technology; 2020.
37. Lackner MA, Rotea MA. Structural control of floating wind turbines. Mechatronics. 2011;21(4):704–19.
38. Carcangiu C E, Pineda I, Fischer T, et al. Wind turbine structural damping control for tower load reduction. In: Civil engineering topics, vol 4. Springer New York; 2011, pp 141–153.
39. Wu JL. Design of offshore wind turbine. Beijing: Chemical Industry Press; 2012.
40. Zeng W. Anti-corrosion design of offshore wind power. Total Corrosion Control. 2020;34(06):97–102.
41. Wang FP, Jing HM, Xin CM. Corrosion electrochemistry (2nd edition). Beijing: Chemical Industry Press; 2017.
42. Zhao XD, Yang J, Zhang Q, et al. Marine corrosion and biological fouling protection technology. Wuhan: Huazhong University of Science and Technology Press; 2017.
43. Ma AB, Jiang JH, et al. Anticorrosion engineering of offshore wind farm. Beijing: China Water Resources and Hydropower Press; 2015.
44. Hou BR. Corrosion control technology of ocean steel structure in spray splash zone. Beijing: Science Press; 2011.
45. Wang T, Li P, Ni Y, et al. Corrosion protection of marine energy buildings. Total Corrosion Control. 2013;06:7–8.
46. Yuan Y, Yan HM, Zhang G, et al. Offshore wind power generation technology. Nanjing: Hohai University Press; 2014.
47. Certification specification for offshore wind turbines. Beijing: China Classification Society, 2012.
48. Energy Industry Wind Power Standardization Technical Committee. Technical specification for lightning protection system of wind turbine: NB/T 31039-2012. Beijing: China Electric Power Press; 2013.

Chapter 4
Offshore Wind Turbine Foundation Structure

The surrounding marine ecology is impacted by offshore wind installations over their whole existence. These consequences include chemical and biological effects on the environment and marine life in addition to physical ones like noise and electromagnetic waves. As a result, this chapter explores the whole range of environmental impacts—physical, chemical, and biological—that offshore wind energy has to offer. Additionally, it presents innovative integrated models for the growth of offshore wind energy, including wave energy, marine ranching, hydrogen production and storage, and other power generation methods.

As a significant load-bearing component of offshore wind turbines, the structural reaction characteristics of the foundation under operating conditions are related to the overall operational safety of the wind turbines. Accurately establishing the structural reaction characteristics is particularly crucial for optimizing the foundation construction. The impact of different offshore wind turbine foundation constructions on the overall dynamic properties, total weight, and applicable environment of wind turbines varies.

This chapter will list various types of offshore wind turbine foundation structures, analyze their respective advantages and disadvantages, propose dynamic design methods and fatigue design methods for foundation structures, provide several common numerical solutions for dynamic equations, and design methods for foundation collision and erosion prevention.

M. Ge et al., *Offshore Wind Turbine Technology*,
https://doi.org/10.1007/978-981-96-2175-0_4

4.1 Types of Foundation Structures for Offshore Wind Turbines

4.1.1 Pile (Bearing) Foundation

Pile (bearing) foundation can be separated into single pile foundation, tripod foundation, high three pile gantry foundation, etc. according to the structural form.

1. **Single pile foundation**

Single pile foundation is the simplest and most generally used foundation structure among pile (bearing) foundations, appropriate for water depths less than 30m and with relatively hard seabed, especially in shallow water near the sea. The diameter of a single pile is normally 3–6 m or even larger depending on the amount of the load.

The connection methods between single pile foundation and tower include welding connection, flange bolt connection, and casing grouting connection. The application of casing grouting connection is the most extensive and the technology is mature. Construction technology of transition section grouting connected single pile foundation is relatively simple and can be summarized as three steps: prefabrication of steel pipe piles and transition sections; Transportation, sinking and piling of steel pipe piles; Installation and grouting of transition section.

According to different geological conditions, there are mainly two methods of pile driving for single pile foundations: after reaching the designated location, the pile hammer is installed on the tubular pile to drive the pile until the pile foundation enters the required seabed depth; Use a drilling machine to drill holes on the seabed, install them into pile foundations, and then pour cement. Structurally speaking, shallow water depth and shallow burial depth of bedrock are the greatest possibilities for single pile foundations, as relatively short rock channels can resist the overturning load of the entire foundation. If the bedrock is buried deep, the pile foundation needs to be driven very deep to resist wind turbine loads and environmental loads. From the perspective of construction difficulty, for hard rocks, especially granite rocks, it is difficult to drive piles on the seabed, requiring drilling and raising expenses. Therefore, driving piles on clay or sandy soil foundations are more convenient.

Due to the extended "free section" of the single pile foundation, the collective stiffness is modest, and the dynamic response is considerable, it is easily constrained by geology and water depth constraints. In addition, because to the vulnerability of single pile foundations to erosion, it is required to protect the seafloor around the piles from erosion.

The development of single pile foundation in China was originally constrained due to the huge pile diameter and the need for large-scale pile driving equipment. In recent years, China has introduced massive pile hammers from abroad, which has strengthened its pile drive building capability. At present, single pile foundations have been effectively implemented in offshore wind farms such as Rudong in Jiangsu.

2. **Tripod foundation**

The tripod (three pile pipe rack) foundation originated from offshore oil and gas engineering and has the advantages of light weight and inexpensive price. The structural form of the foundation is: the upper portion of the main cylinder (single column) is welded with three slant support, and the bottom section is welded with three horizontal braces, which are used to bear and transfer the load from the upper tower cylinder. A steel pile is installed at the lowest three legs to fix the foundation. The three pile tremie is pushed into the seabed to a specified depth, and the pile top supports the higher tower cylinder through the steel casing, such that the tripod and the pile form a joint foundation. The appropriate water depth is 20–80 m. Tripod foundations are currently extensively employed in offshore wind farms in Germany.

3. **High grade three pile gantry foundation**

Similar to a tripod foundation, the High three pile gantry foundation includes putting three supporting piles into the bottom of the seafloor, with the piles and the portal connecting beam is utilized for the upper wind turbine tower instead of the typical slant support plus transverse bracing. The portal connection beam can be separated into two pieces, namely the horizontal transition section near the tower side and the diagonal transition section near the casing side. The foundation of the three piles portal frame has the benefits of good overall integrity, relatively high rigidity, upper casing and lower connection through water grouting, and convenient construction circumstances. Compared with the foundation construction of the three piles jacket, the manufacture of the inclined transition section of the three piles portal frame foundation is more difficult and the pile diameter is bigger.

4. **Other single column multi pile foundations**

With the increase of single machine capacity and wind turbine load, the tripod foundation has evolved from a three-pile jacket to five-pile jacket and six-pile jacket, with significantly improved bearing capacity, increased stiffness, smaller pile diameter than the tripod foundation, and suitable for offshore wind farms with water depths exceeding 30 m.

4.1.2 Gravity Foundation

As the name suggests, a gravity foundation relies on its own weight to resist the loads of wind turbines and environmental loads, and can preserve the foundation's resistance to overturning and sliding.

Gravity based foundations are generally reinforced concrete buildings that require ballast materials. Depending on local conditions, more economical ballast materials such as sand, gravel, slag, and concrete might be utilized. Gravity based foundations have a small bearing capacity, simple structure and manufacturing process, mature application expertise, and relatively low cost. They are the greatest in volume and mass among all types of foundations, and have the best stability and reliability. They offer strong resistance to storms and wind waves, and are appropriate for wind farms with good natural foundations and water depths within 30 m. They are not appropriate for soft soil foundations and washed seabed locations.

Before the construction of gravity foundation, the mud layer on the seabed surface should be cleared. After the excavation of the foundation trench is completed, stones should be promptly thrown and compacted layer by layer to eliminate or reduce compression settlement; When backfilling, attention should be paid to ensuring uniformity in all directions to avoid foundation tilting and cracking of partition walls. A layer of filter layer should be made with crushed stones and geotextile on the top, and concrete should be poured to seal the compartment. Tower connecting rods, flanges, etc. should be pre embedded. Gravity based foundations are generally transported on land using airbags and winches, and on sea using barges, semi-submersible barges, and floating cranes. They can also be transported on their own ability to float and is equipped with a certain airbag to suspend in the water, which is then towed to the installation site by a tugboat.

4.1.3 Truss Type Jacket Foundation

The truss type jacket foundation is a steel cone platform space frame (truss) structure, which has the advantages of high overall structural stiffness and light weight due to its geometric properties. Can normally be designed with 3 or 4 piles. Foreign wind farms often utilize a 4-pile duct frame structure.

During the construction of the truss type jacket foundation, the steel pipe skeleton welding work needs to be done on land first, then moved to the selected installation location, fixed by pile driving operations, and lastly installed with tower and wind turbine units.

The support pipe of the conduit rack has X-shaped support, K-shaped support, and monoclinic support. The vertical support pipe is a twin inclined space steel pipe. According to the manner of fixing the jacket with foundation piles, it can be classified into main pile type fixed jacket and skirt pile type uniformly fastened around the bottom of the jacket.

4.1.4 Pile Group Cap Foundation

Multi pile foundation can be split into high pile foundation and low pile foundation according to the height of the foundation. Due to the usage of cast-in-place concrete abutments in the upper half, the foundation structure is rather heavy and the rigidity of the abutments itself is high. To improve the stiffness of the lowest section of the foundation, a particular length of concrete is poured inside the steel pipe piles below the bearing platform. Multiple foundation piles are embedded in the concrete bearing platform for a given length through pouring. From the perspective of structural stress and control of horizontal displacement, the foundation piles are commonly constructed as inclined piles.

1. **High pile cap foundation**

During construction, after the pile foundation of the high-pile cap is driven, steel box templates are installed at the top or side of the pile using auxiliary facilities such as pile clamps, support beams, and bottom steel plates. Steel casing can serve as a water barrier. With the completion of cross sea bridges such as Donghai Bridge and Hangzhou Bay Bridge, relevant offshore construction units have accumulated rich experience in offshore pier and abutment structure construction, and the construction technology is relatively mature. The high pile cap foundation is a commonly used structure for coastal docks and bridge piers, consisting of pile foundations and caps. Currently, it has not been applied in offshore wind turbine foundations abroad. The first offshore wind farm in China, the Shanghai Donghai Bridge offshore wind farm, and the Jiangsu Xiangshui test wind turbine unit, adopted this foundation structure. Compared to low pile foundation, high pile foundation has a higher bottom height, and the construction foundation of the foundation is not affected by waves, with a higher center of gravity.

2. **Low pile cap foundation**

The low pile cap foundation forms a seawater free hollow environment due to the need for steel boxed boxes, which is suited for the development of intertidal marine areas and mudflat wind farms. The bottom elevation of the low pile foundation is set at the average sea level, and the building of the foundation is to some extent affected by tides and waves. The center of gravity of the foundation is low, the overall stiffness is high, and the anti-collision ability is good. For intertidal zone and mudflat region, because the seabed will be exposed to the water after the ebb tide, which can generate a land building environment, it is more suitable for the use of low pile cap foundation. The low pile foundation has been successfully implemented to the intertidal wind farm in Rudong, Jiangsu.

4.1.5 Bucket Foundation

Bucket foundation (negative pressure bucket foundation) can be split into single bucket, three bucket, and multiple bucket kinds, and can be deployed in both shallow and deep waters. Among them, the negative pressure bucket foundation in shallow waters is actually a combination of traditional pile foundations and gravity foundations; The negative pressure bucket foundation in the deep sea, as an anchoring system for tension leg floating support, can better reflect the economic advantages. Bucket foundations are currently mostly employed in shallow waterways.

The bucket foundation structure has inverted open circular buckets, which can be prefabricated with steel or reinforced concrete and belong to short and thick rigid piles. Each bucket is attached to the steel circular bucket by a central column and a shear plate with reinforcing bars. The shear plate distributes the load of the central column to the bucket wall and transmits it to the foundation. During construction, the prefabricated steel drums on land are placed in water and inflated, then floated to the intended spot. The ventilation holes on the top of the drums are opened to discharge the gas inside the drums. At this time, seawater rushes in and the drums sink to the bottom or mud surface. After sinking to the seabed or mud surface, gas and water are drained from the ventilation holes on the top of the drums to create vacuum pressure and pressure differential inside and outside the drums. Using this negative pressure action, the drums are inserted. The seabed has a set depth to achieve a fixed effect.

As a new form of wind turbine foundation structure, bucket foundation has the key benefits of convenient construction, negative pressure concept for bucket sinking, and no requirement for pile driving equipment and large cranes. Compared with single pile foundations, bucket foundations offer higher rigidity, lower dynamic response, no noise during construction, can be carried by floating, are easy to deconstruct, and can be readily lifted by balancing the pressure inside and outside the caisson.

The primary constraints of adopting bucket foundations are: complex manufacturing processes and extra procedures compared to single pile foundations; The requirements for topography and geology are rather high, which can only be utilized for soft soil foundation, not for scouring seabed, lithologic seabed, compressible silt, and mudflat and intertidal zone that can't ensure that the foundation is always submerged; Shipping by sea requires a particular depth of water at the port.

In summary, the main advantage of bucket foundation is in its convenient construction, however due to the participation of negative pressure penetration principle, there are many aspects considered in the design, making the design complicated and resulting in considerable investment fluctuations. In addition, during construction, tilting should be controlled throughout the sinking process, and strict requirements are made on negative pressure penetration. The water pressure difference within and outside the bucket created by negative pressure might cause soil seepage. Although it can substantially reduce sinking resistance, excessive seepage will cause large deformation of the soil inside the bucket, generating soil plugs, and even potentially liquefying and flowing the dirt inside the bucket. Bucket foundation started rather late and has not been developed for a long time. Although it is widely utilized in offshore port engineering in China, its application in offshore wind turbine foundations is not yet mature enough, and some risk analyses are not complete. Therefore, it has not been thoroughly utilized and is still in the research and experimental stage.

4.1.6 Floating Foundations

Compared to nearshore areas, the wind energy resources in deep-sea areas are more abundant. When developing offshore wind farms in 50m or deeper sea areas, employing typical fixed pile foundations or duct frame foundations would be costly and unable to expand to deeper sea areas. With the advancement of offshore platform technology, the notion of floating foundation provides the ability for offshore wind power to extend towards the deep sea.

Floating foundations include Tension Leg Platform (TLP) floating foundations, Spar Buoy floating foundations, Tri-floater floating foundations, and various novel types of floating foundations Floating foundation, etc.

- The tension leg floating foundation essentially fixes and maintains the complete wind turbine by the tension of mooring ropes to preserve stability.
- The cylindrical floating foundation is attached to the entire wind turbine unit through three catenary wires, and the ballast tank presses the center of gravity of the entire system below the floating center to ensure the stability of the wind turbine unit in water.
- The three floating box floating foundation principally relies on the catenary to fix the complete wind power foundation, and maintains stability by balancing the gravity and buoyancy of each floating box individually.

Compared to fixed foundations, floating foundations serve as platforms for placing wind turbines and are anchored to the seafloor utilizing anchoring mechanisms. They are relatively low-cost and easy to transport. However, because to their weak stability, they are strongly affected by external elements such as sea winds, waves, and currents, making the design of the platform and anchoring mechanism hard. Therefore, they must have adequate buoyancy to sustain the wind turbines and be able to suppress tilting, shaking, and normal movement within acceptable limits to ensure their regular operation.

In summary, although the construction of offshore wind farms in China started relatively late, with the enhancement of China's comprehensive national strength, the improvement of economic and scientific research capabilities, and the increase of policy support, research on offshore wind farm construction and related fields will usher in a period of rapid development. While drawing on foreign design, construction and other methods and experiences, combined with China's national conditions, we will definitely explore offshore wind turbine foundation structures that are suitable for the environmental conditions, construction conditions and technological strength of China's offshore wind farms.

4.2 Offshore Wind Turbine Foundation Design

4.2.1 Dynamics Design

Combine the finite element models of wind turbines, supporting structures, and foundation components into a dynamic equation that may be integrated for analysis. When it is a linear model, the dynamic equation is [1]

$$\boldsymbol{M}\ddot{\boldsymbol{u}}(t) + \boldsymbol{C}\dot{\boldsymbol{u}}(t) + \boldsymbol{K}\boldsymbol{u}(t) = \boldsymbol{F}(t) \tag{4.1}$$

In the formula, $\boldsymbol{M}$, $\boldsymbol{C}$ and $\boldsymbol{K}$ represent the overall mass matrix; damping matrix and stiffness matrix, respectively; $\ddot{\boldsymbol{u}}(t)$, $\dot{\boldsymbol{u}}(t)$ and $\boldsymbol{u}(t)$ represent the acceleration, the velocity and the displacement at time t, respectively.

When it is a nonlinear model, the stiffness matrix nonlinear dynamic equation can be formulated as follows.

$$\boldsymbol{M}\ddot{\boldsymbol{u}}(t) + \boldsymbol{C}\dot{\boldsymbol{u}}(t) + \boldsymbol{K}[\boldsymbol{u}(t)]\boldsymbol{u}(t) = \boldsymbol{F}(t) \tag{4.2}$$

Equations (4.1) and (4.2) are second-order partial differential equations involving structural displacement, which need to be solved for initial and boundary conditions.

4.2.2 Numerical Solution of Dynamic Equations

The numerical methods for solving linear dynamic equations include central difference method, Wilson-θ method, and Newmark-β method.

1. **Central difference method**

If the system states $\boldsymbol{u}(t)$ and $\dot{\boldsymbol{u}}(t)$ at time t are known, then the motion of the system at the next time $t + \Delta t$ satisfies:

$$\boldsymbol{M}\ddot{\boldsymbol{u}}(t+\Delta t) + \boldsymbol{C}\dot{\boldsymbol{u}}(t+\Delta t) + \boldsymbol{K}\boldsymbol{u}(t+\Delta t) = \boldsymbol{F}(t+\Delta t) \tag{4.3}$$

For a sufficiently small interval Δt, the Taylor expansion can be written as follows:

$$\begin{cases} \boldsymbol{u}(t+\Delta t) = \boldsymbol{u}(t) + \Delta t\dot{\boldsymbol{u}}(t) + \dfrac{\Delta t^2}{2}\ddot{\boldsymbol{u}}(t) + \dfrac{\Delta t^3}{6}\dddot{\boldsymbol{u}}(t) + \cdots \\ \dot{\boldsymbol{u}}(t+\Delta t) = \dot{\boldsymbol{u}}(t) + \Delta t\ddot{\boldsymbol{u}}(t) + \dfrac{\Delta t^2}{2}\dddot{\boldsymbol{u}}(t) + \cdots \\ \ddot{\boldsymbol{u}}(t+\Delta t) = \ddot{\boldsymbol{u}}(t) + \Delta t\dddot{\boldsymbol{u}}(t) + \cdots \end{cases} \tag{4.4}$$

If the high-order terms of Taylor expansion mentioned above are ignored, the acceleration varies linearly with time within $[t, t + \Delta t]$. This implies the assumption of linear variation of external excitation, which requires a sufficiently small interval Δt to approximate the true external excitation.

Encouragement Eq. (4.4) contains four unknown vectors that satisfy the necessary conditions for solving. The specific solution can be written as:

$$\dddot{\boldsymbol{u}}(t) = \frac{6}{\Delta t^3}\left[\boldsymbol{u}(t+\Delta t) - \boldsymbol{u}(t) - \Delta t\dot{\boldsymbol{u}}(t) - \frac{1}{2}\Delta t^2\ddot{\boldsymbol{u}}(t)\right] \tag{4.5}$$

$$\begin{cases} \dot{\boldsymbol{u}}(t+\Delta t) = \dfrac{3}{\Delta t}\boldsymbol{u}(t+\Delta t) - \boldsymbol{\alpha}_1(t) \\ \ddot{\boldsymbol{u}}(t+\Delta t) = \dfrac{6}{\Delta t^2}\boldsymbol{u}(t+\Delta t) - \boldsymbol{\alpha}_2(t) \end{cases} \tag{4.6}$$

In the formula,

$$\begin{cases} \boldsymbol{\alpha}_1(t) = \dfrac{3}{\Delta t}\boldsymbol{u}(t) + 2\dot{\boldsymbol{u}}(t) + \dfrac{\Delta t}{2}\ddot{\boldsymbol{u}}(t) \\ \boldsymbol{\alpha}_2(t) = \dfrac{6}{\Delta t^2}\boldsymbol{u}(t) + \dfrac{6}{\Delta t}\dot{\boldsymbol{u}}(t) + 2\ddot{\boldsymbol{u}}(t) \end{cases} \tag{4.7}$$

By substituting the Formula (4.7) into (4.3), it yields:

$$\left(\frac{6}{\Delta t^2}\boldsymbol{M} + \frac{3}{\Delta t}\boldsymbol{C} + \boldsymbol{K}\right)\boldsymbol{u}(t+\Delta t) = \boldsymbol{F}(t+\Delta t) + \boldsymbol{M}\boldsymbol{\alpha}_2(t) + \boldsymbol{C}\boldsymbol{\alpha}_1(t) \tag{4.8}$$

From this, it can be inferred that

$$\boldsymbol{u}(t+\Delta t) = \left(\frac{6}{\Delta t^2}\boldsymbol{M} + \frac{3}{\Delta t}\boldsymbol{C} + \boldsymbol{K}\right)^{-1}[\boldsymbol{F}(t+\Delta t) + \boldsymbol{M}\boldsymbol{\alpha}_2(t) + \boldsymbol{C}\boldsymbol{\alpha}_1(t)] \tag{4.9}$$

A notable issue with the above strategy is the discrepancy between computational precision and computation time. To ensure calculation accuracy, the time step Δt should be suitably small. A smaller Δt will increase the number of recursive steps and lengthen the calculation time. In addition, increasing the number of recursive stages will likewise raise the cumulative error. Therefore, one of the essential criteria for evaluating the direct integration method is to allow the use of the maximum integration step size. How an algorithm does not diverge at any step size, and is deemed unconditionally stable. If the solution does not diverge only within a given step size range, it is called conditionally stable.

2. **Wilson-θ method**

This method is based on another form of expansion of $\boldsymbol{u}(t + \Delta\, t)$, if introduced:

$$\Delta\ddot{\boldsymbol{u}}_s(t) = \ddot{\boldsymbol{u}}(t+s) - \ddot{\boldsymbol{u}}(t) \tag{4.10}$$

Then there are:

$$\Delta\ddot{\boldsymbol{u}}_s(t+\tau) = \ddot{\boldsymbol{u}}(t) + \frac{\tau}{s}\ddot{\boldsymbol{u}}_s(t) \quad (0 \le \tau \le s) \tag{4.11}$$

Integrating τ once and twice, let $\tau = s = \theta\Delta t$, we can obtain:

$$\begin{cases} \dot{\boldsymbol{u}}_s(t+s) = \dot{\boldsymbol{u}}(t) + s\ddot{\boldsymbol{u}}(t) + \dfrac{s}{2}\ddot{\boldsymbol{u}}_s(t) \\ \boldsymbol{u}(t+s) = \boldsymbol{u}(t) + s\boldsymbol{u}(t) + \dfrac{s}{2}\ddot{\boldsymbol{u}}(t) + \dfrac{s^2}{6}\Delta\ddot{\boldsymbol{u}}_s(t) \end{cases} \tag{4.12}$$

$$\begin{cases} \boldsymbol{u}_s(t+s) = \dot{\boldsymbol{u}}(t) + \dfrac{s}{2}[\ddot{\boldsymbol{u}}_s(t+s) + \ddot{\boldsymbol{u}}(t)] \\ \boldsymbol{u}(t+s) = \boldsymbol{u}(t) + s\dot{\boldsymbol{u}}(t) + \dfrac{s^2}{6}[\ddot{\boldsymbol{u}}(t+s) + 2\ddot{\boldsymbol{u}}(t)] \end{cases} \tag{4.13}$$

$$\begin{cases} \ddot{\boldsymbol{u}}(t+s) = \dfrac{6}{s^2}[\boldsymbol{u}(t+s) - \boldsymbol{u}(t)] - \dfrac{6}{s}\dot{\boldsymbol{u}}(t) - 2\ddot{\boldsymbol{u}}(t) \\ \dot{\boldsymbol{u}}(t+s) = \dfrac{3}{s}[\boldsymbol{u}(t+s) - \boldsymbol{u}(t)] - 2\dot{\boldsymbol{u}}(t) - \dfrac{s}{2}\ddot{\boldsymbol{u}}(t) \end{cases} \tag{4.14}$$

Write down the differential equation of motion for the system at time $t + s$ as follows:

$$\boldsymbol{M}\ddot{\boldsymbol{u}}(t+s) + \boldsymbol{C}\dot{\boldsymbol{u}}(t+s) + \boldsymbol{K}\boldsymbol{u}(t+s) = \boldsymbol{F}(t+s) \tag{4.15}$$

The linear algebraic equation that the system displacement should satisfy at time $t + s$ is:

$$\tilde{\boldsymbol{K}}(t)\boldsymbol{u}(t+s) = \boldsymbol{g}(t, t+s) \tag{4.16}$$

$$\tilde{\boldsymbol{K}} = \frac{6}{s^2}\boldsymbol{M} + \frac{3}{s}\boldsymbol{C} + \boldsymbol{K} \tag{4.17}$$

$$\begin{aligned} g(t, t+s) = \boldsymbol{M}\left[\frac{6}{s^2}\boldsymbol{u}(t) + \frac{6}{s}\dot{\boldsymbol{u}}(t) + 2\ddot{\boldsymbol{u}}(t)\right] + \boldsymbol{C}\left[\frac{3}{s}\boldsymbol{u}(t) + 2\dot{\boldsymbol{u}}(t) + \frac{s}{2}\ddot{\boldsymbol{u}}(t)\right] \\ + F(t+s) \end{aligned} \tag{4.18}$$

Using linear extrapolation, i.e.:

$$\boldsymbol{F}(t+s) = \boldsymbol{F}(t) + s[\boldsymbol{F}(t+\Delta t) - \boldsymbol{F}(t)] \tag{4.19}$$

Ultimately, it can be obtained:

$$\begin{cases} \ddot{\boldsymbol{u}}(t, t+\Delta t) = \dfrac{6}{\theta^2\Delta t^2}[\boldsymbol{u}(t+s) - \boldsymbol{u}(t)] - \dfrac{6}{\theta^2\Delta t}\dot{\boldsymbol{u}}(t) + \left(1 - \dfrac{3}{\theta}\right)\ddot{\boldsymbol{u}}(t) \\ \dot{\boldsymbol{u}}(t, t+\Delta t) = \dot{\boldsymbol{u}}(t) + \dfrac{\Delta t}{2}[\ddot{\boldsymbol{u}}(t+s) + \ddot{\boldsymbol{u}}(t)] \\ \boldsymbol{u}(t, t+\Delta t) = \boldsymbol{u}(t) + \Delta t\dot{\boldsymbol{u}}(t) + \dfrac{\Delta t^3}{6}[\ddot{\boldsymbol{u}}(t+s) + 2\ddot{\boldsymbol{u}}(t)] \end{cases} \tag{4.20}$$

If the Wilson-θ method is used to solve Eq. (4.1), the specific steps are as follows.

- Form overall matrices K, M, and C based on initial conditions.
- Given the initial displacement, velocity, and acceleration values $\ddot{\boldsymbol{u}}(0)$, $\dot{\boldsymbol{u}}(0)$, and $\boldsymbol{u}(0)$.
- Select the time step Δt and parameter δ, and specify the integration parameter θ = 1.4. Calculate the integral constant b_i.

$$
\begin{aligned}
& b_0 \geq \frac{6}{(\theta \Delta t)^2}, b_1 \geq \frac{3}{\theta \Delta t}, b_2 = 2b_1 \\
& b_3 = \frac{\theta \Delta t}{2}, b_4 = \frac{b_0}{\theta}, b_5 = -\frac{b_2}{\theta} \\
& b_6 = 1 - \frac{3}{\theta}, b_7 = \frac{\Delta t}{2}, b_8 = \frac{\Delta t^2}{6}
\end{aligned}
\tag{4.21}
$$

- Form an effective stiffness matrix, i.e.

$$
\hat{\boldsymbol{K}} = \boldsymbol{K} + b_0 \boldsymbol{M} + b_1 \boldsymbol{C} \tag{4.22}
$$

- For each time step (t = 0, Δt, 2Δt), there are

$$
\hat{\boldsymbol{K}} \boldsymbol{u}(t + \theta \Delta t) = \boldsymbol{F}(t + \theta \Delta t) \tag{4.23}
$$

Obtain the displacement, velocity, and acceleration at time $t + \Delta t$, i.e.

$$
\boldsymbol{u}(t + \Delta t) = \boldsymbol{u}(t) + \Delta t \dot{\boldsymbol{u}}(t) + b_8[\ddot{\boldsymbol{u}}(t + \theta \Delta t) + 2\ddot{\boldsymbol{u}}(t)] \tag{4.24}
$$

$$
\dot{\boldsymbol{u}}(t + \Delta t) = \dot{\boldsymbol{u}}(t) + b_7[\ddot{\boldsymbol{u}}(t + \theta \Delta t) + \ddot{\boldsymbol{u}}(t)] \tag{4.25}
$$

$$
\ddot{\boldsymbol{u}}(t + \Delta t) = b_4[\boldsymbol{u}(t + \theta \Delta t) - \boldsymbol{u}(t)] + b_5 \dot{\boldsymbol{u}}(t) + b_6 \ddot{\boldsymbol{u}}(t) \tag{4.26}
$$

3. **Newmark-$\boldsymbol{\beta}$ method**

If the Newmark integration method is used to solve Eq. (4.1), the specific steps are as follows:

- Form overall matrices K, M, and C based on initial conditions.
- Given the initial displacement, velocity, and acceleration values, namely $\ddot{\boldsymbol{u}}(0)$, $\dot{\boldsymbol{u}}(0)$, and $\boldsymbol{u}(0)$.
- Select the time step Δt and parameter δ, calculate the value of parameter α, and define the common Newmark integral constant as:

$$\delta \geq 0.5, \alpha \geq 0.25(0.5+\delta)^2$$
$$c_0 = \frac{1}{\alpha \Delta t^2}, c_1 = \frac{\delta}{\alpha \Delta t}, c_2 = \frac{1}{\alpha \Delta t}, c_3 = \frac{1}{2\alpha} - 1$$
$$c_4 = \frac{\delta}{\alpha} - 1, c_5 = \frac{\Delta t}{2}\left(\frac{\delta}{\alpha} - 2\right), c_6 = \Delta t(1-\delta), c_7 = \Delta t\delta \tag{4.27}$$

- Form an $\hat{\boldsymbol{K}}$ effective stiffness matrix, i.e.

$$\hat{\boldsymbol{K}} = \boldsymbol{K} + c_0\boldsymbol{M} + \boldsymbol{c}_1\boldsymbol{C} \tag{4.28}$$

- For each time step (t = 0, Δt, 2Δt), calculate the payload at time t + Δt, i.e.

$$\hat{\boldsymbol{Q}}(t+\Delta t) = \boldsymbol{Q}(t+\Delta t) + \boldsymbol{M}(\boldsymbol{c}_0\boldsymbol{u}_t + \boldsymbol{c}_2\dot{\boldsymbol{u}}_t + \boldsymbol{c}_3\ddot{\boldsymbol{u}}_t) + \boldsymbol{C}(\boldsymbol{c}_1\boldsymbol{u}_t + \boldsymbol{c}_4\dot{\boldsymbol{u}}_t + \boldsymbol{c}_5\ddot{\boldsymbol{u}}_t) \tag{4.29}$$

- For each time step (tt = 0,Δt,2Δt), solve for the displacement at time t + Δt, i.e.

$$\hat{\boldsymbol{K}}u(t+\Delta t) = \boldsymbol{Q}(t+\Delta t) \tag{4.30}$$

- For each time step(t = 0, Δt, 2Δt), calculate the velocity and acceleration at time t + Δt

$$\ddot{\boldsymbol{u}}(t+\Delta t) = c_0(\boldsymbol{u}(t+\Delta t) - \boldsymbol{u}(t)) - c_2\dot{\boldsymbol{u}}(t) - c_3\ddot{\boldsymbol{u}}(t) \tag{4.31}$$

4.2.3 Fatigue Design

1. **Basic concepts of fatigue**

The American Society for Testing and Materials (ASTM) defines fatigue as the process of localized permanent structural changes occurring in a material subjected to conditions which procedure fluctuating stress and strains at some point or points and which may culminate in crack or complete fracture after a sufficient number of fluctuations.

The simplest cyclic load is a constant amplitude cyclic stress load. Sinusoidal constant amplitude cyclic stress is exhibited in Fig. 4.1 as indicated. Obviously, representing a cyclic stress requires at least two numbers, such as the maximum stress Smax and the minimum stress S_{min}, which are the fundamental quantities for describing the amount of cyclic stress.

When conducting fatigue analysis, the following parameters are commonly used, namely stress range (full amplitude) ΔS, defined as

$$\Delta S = S_{\max} - S_{\min} \tag{4.32}$$

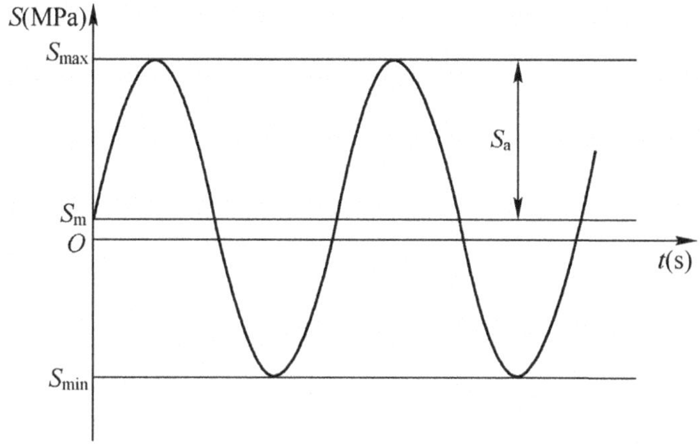

Fig. 4.1 Sinusoidal constant amplitude cyclic stress

The stress amplitude (half amplitude) S_a is defined as

$$S_a = \Delta S/2 = (S_{\max} - S_{\min})/2 \tag{4.33}$$

The mean stress S_m is defined as

$$S_m = \Delta S/2 = (S_{\max} + S_{\min})/2 \tag{4.34}$$

The stress ratio R is defined as

$$R = S_{\min}/S_{\max} \tag{4.35}$$

Among them, the stress ratio R reflects different cycle characteristics, for example, when $S_{\max} = -S_{\min}$, $R = -1$, it is a symmetric cycle. When $S_{\min} = 0$ and $R = 0$, it is pulse cycle. When $S_{\max} = S_{\min}$, $R = 1$, $S_a = 0$, the static load. In the above parameters, only two of them need to be known to determine the cyclic stress level.

2. **Stress fatigue**

Fatigue can be separated into stress fatigue and strain fatigue depending to the level of the cyclic stress applied. If the maximum cyclic stress is substantially lower than the yield stress, it is called stress fatigue. Due to the low degree of cyclic stress applied, the number of lifecycles is relatively high (usually greater than 104), hence it is also known as high cycle fatigue. If the maximum cyclic stress is more than the yield stress, then due to the large strain shift after the material yields, the stress change is relatively minor, and utilizing strain as a fatigue control parameter is more appropriate, consequently it is called strain fatigue. Due to the high degree of cyclic stress induced by strain fatigue, the lifespan is relatively low, and the number of life cycles is often fewer than 104. Strain fatigue is also known as low cycle fatigue.

3. **S-N curve**

The S–N curve is a description of the relationship between the stress S acting on the fatigue performance of a material and the life N at failure.

The simplest load spectrum under fatigue loading is constant amplitude cyclic stress loading. Describing the level of cyclic stress requires two quantities, namely stress ratio R and stress amplitude S. The stress ratio determines the cyclic characteristics, and the stress amplitude is the controlling parameter for fatigue failure.

When R = −1, under the control of symmetric constant amplitude cyclic stress load, the stress–strain relationship given by the experiment is expressed by the Sa-N curve, which is the basic fatigue performance curve of the material. When R = −1, Sa = Smax, so the basic stress life curve is also known as the S–N curve. The stress S can be Sa or Smax, the two values are equal.

Fatigue failure has three stages: crack initiation, stable propagation, and unstable propagation to fracture. This study focuses on the crack initiation stage. Therefore, destruction can be defined as:

- The standard small-sized specimen fractures. For brittle materials such as high-strength steel and medium strong steel, the time from crack initiation to propagation to the fracture of small-sized circular section specimens is very short, and the impact on the entire life is small. Considering the small size of crack initiation and the difficulty of observation, this definition is reasonable.
- Visible small cracks (such as 1 mm) or a strain drop of 5–15% appear. For materials with good ductility, there is a considerable extension stage after crack initiation, which should not be included in the crack initiation stage. When it is difficult to observe small-sized cracks, the strain changes under constant amplitude cyclic stress can be monitored. When cracks appear in the specimen, the stiffness changes and the strain also changes accordingly. Therefore, the strain change can be used to determine whether cracks have occurred.

The standard specimens used for fatigue performance testing of materials are generally small-sized smooth cylindrical specimens with a diameter of 3 to 10 mm. The basic *S–N* curve represents the crack initiation life of smooth materials under constant amplitude symmetrical cyclic stress.

To obtain the *S–N* curve at a given stress ratio R, a set of standard specimens (typically 7 to 10 pieces) is subjected to different stress amplitudes S_a during fatigue testing, and the corresponding lifetimes N are recorded. This allows for the construction of the *S–N* curve, reflecting the fatigue performance of the material at various stress amplitudes.

4. **Influence of average stress**

The *S–N* curve reflecting the fatigue performance of materials is obtained under a given stress ratio R. When $R = -1$, the *S–N* curve during symmetric cycling is the basic S–N curve.

The increase in the tensile part of cyclic loading is detrimental to the initiation and propagation of fatigue cracks,

Can reduce fatigue life. The general trend of the average stress on the *S–N* curve is that when the average stress $S_m = 0$ ($R = -1$), the *S–N* curve is the basic *S–N* curve; When $S_m > 0$, that is, when the average tensile stress is applied, the *S–N* curve decreases, or in other words, the fatigue strength decreases under the same fatigue life, which has an adverse effect on fatigue; When $S_m < 0$, that is, when the average compressive stress is applied, the *S–N* curve shifts upward, indicating that under the same pressure amplitude.

The increase in service life, or in other words, the increase in fatigue strength under the same service life, therefore the influence of compressive average stress on fatigue is beneficial.

5. **Palmgrem-Miner theory**

If the lifespan of a component under a constant amplitude cyclic stress *S* is *N*, then the damage after n cycles can be defined as

$$D = n/N \tag{4.36}$$

Obviously, under constant amplitude cyclic stress *S*, if $n = 0$, then $D = 0$, and the component is not subjected to fatigue damage; If $n = N$, then $D = 1$, the component experiences fatigue damage.

The damage suffered by the component under constant amplitude cyclic stress S_i after n_i cycles is $D_i = n_i/N_i$. If under the action of *k* constant amplitude cyclic stresses S_i, each subjected to n_i cycles, the total damage is defined as

$$D = \sum_{1}^{k} D_i = \sum n_i/D_i \tag{4.37}$$

The criteria for destruction are

$$D = \sum n_i/N_i = 1 \tag{4.38}$$

This is the simplest, most famous, and widely used Miner's linear cumulative damage theory. Among them, n_i is the number of cycles under the action of S_i, given by the load spectrum; N_i is cycled to failure under the action of S_i. The lifespan is determined by the *S–N* curve.

As mentioned earlier, the total damage $D = 1$ is the empirical failure criterion for Miner's linear cumulative damage. Considering the influence of spectral order, it can actually be

$$D = \sum n_i/N_i = Q \tag{4.39}$$

In the formula, Q is related to the load spectrum, the order of action, and the dispersion of the material. In fact, for a specific component, the value of Q can be determined by drawing on past usage experience or experimental data of similar components, and the estimated Q can reflect the influence of actual load sequence, etc.

6. **Random load spectrum and cycle counting method**

Under constant amplitude load, the fatigue life can be directly estimated using the S–N curve. For variable amplitude loads' life prediction, it can also be solved with the help of Miner theory under load spectrum. If the random load spectrum can be equivalently converted into a variable amplitude or constant amplitude load spectrum, previous methods can be used to analyze fatigue problems.

(1) Random load spectrum and several definitions

Figure 4.2 illustrates the random load spectrum, representing a scenario in which the load fluctuates unpredictably over time, referred to as the load time history. Random load spectra are typically derived from actual measurements of standard operating circumstances. Discussing the use of computers is prior to converting a random load spectrum into a variable amplitude load spectrum through numerical method, let me introduce the relevant definitions of the fatigue analysis cyclic counting method (ASTM E1049-85).

- Load represents parameters including force, stress, strain, displacement, torque, acceleration, and other associated parameters.
- The points on the reversed load-time history curve where the slope changes sign are as follows: The point where the slope changes from positive to negative is referred to as a "peak," while the point where the slope changes from negative to positive is called a "trough." Both peaks and troughs are considered reversal points. In a constant amplitude cycle spectrum, each cycle contains two reversals.

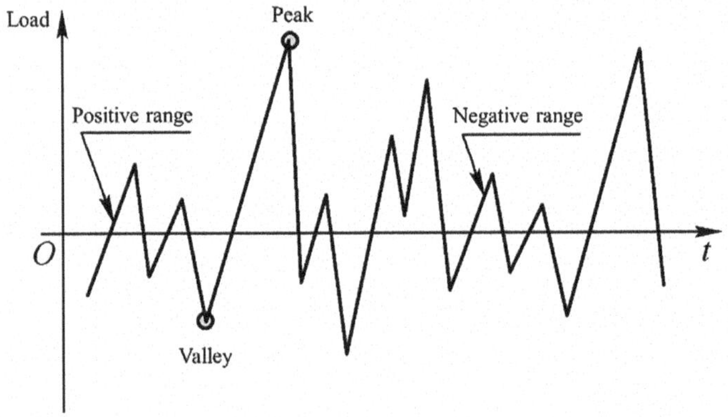

Fig. 4.2 Random load spectrum

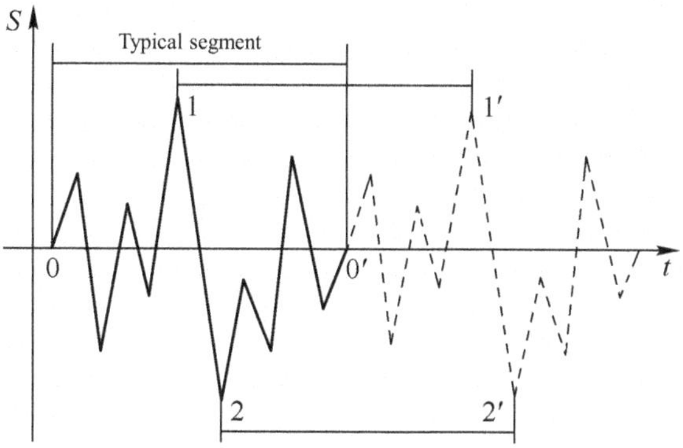

Fig. 4.3 Selection of typical segments for simplified rainflow counting method

- Range is the difference between consecutive peak and valley load values: the range between the valley and the succeeding peak loads, characterized by a positive slope, is called positive range; Conversely, the range from peak to subsequent valley load with a negative slope is termed negative range.

(2) Simplified rainflow counting method

The approach of translating irregular and random load-time histories into a sequence of cycles is called the Cycle Counting approach. There are various counting methods available. This section will solely address the simplified rainflow counting approach, which is simple, adaptable, and exhibits consistent stress–strain responses under variable amplitude cyclic loads.

The simplified rainflow counting method is appropriate for repetitive processes derived from typical load spectrum segments. Given that the load is a repetition of a typical segment, taking the starting and ending segments at the maximum peak or valley as the typical segment will maintain its generality, as illustrated in Fig. 4.3.

The procedure for simplifying the rainflow counting method is as follows:

Step 1: Select the maximum peak or valley start and end suitable for simplifying the rainflow counting method from the random load spectrum as a typical counting segment. The simplified rainflow counting process is shown in Fig. 4.4. In Fig. 4.3, 1-1′ represents the start and end of the maximum peak, and 2-2′ represents the start and end of the maximum valley.

Step 2: Rotate the process curve by 90° and consider it as a multi-layered roof, assuming that raindrops start to flow downwards along the maximum peak or valley. If there is no roof obstruction, the raindrops will reverse direction and continue to flow to the endpoint. In Fig. 4.4, the raindrops start from point *A*, flow along *AB*, and then fall onto the *CD* roof at point *B*. They continue to stay at point *D*. As there is no roof obstruction, the raindrops reverse direction and flow along *DE* to point *E*, then fall onto roof *JA*′, and end their flow at point *A*′. The path is *ABDEA*′.

Step 3: Record the maximum peak and valley values left by the raindrops as a cycle. In Fig. 4.4, the cycle given by the first flow path is *ADA*′. The cycle parameters

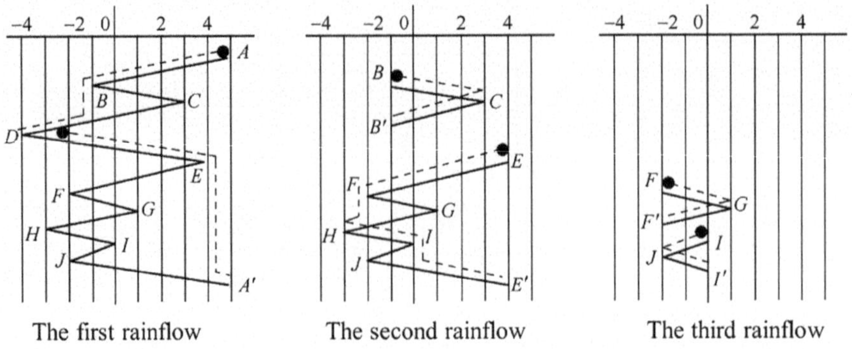

Fig. 4.4 Simplified rainflow counting process

include load range and average load, such as the load range ΔS for the *ADA′* cycle, which is $\Delta S = 5 - (-4) = 9$, and the average load $S_m = (5 - 4)/2 = 0.5$.

Step 4: Remove the part where raindrops flow through the process, and repeat the above rainflow counting for each remaining process segment until there are no more remaining processes. The second rainflow resulted in *BCB′* and *EHE′* cycles. The third rainflow resulted in *FGF′* and *IJI′* cycles. Then counting completed.

The rainflow counting results are presented in Table 4.1. Table 4.1 delineates the loop and its associated parameters. If the load is stress, the range specified in Table 4.1 is ΔS, the stress amplitude $S_a = \Delta S/2$, and the average stress S_m is the mean value in the table. Rain flow counting involves a two parameters count. The loop is entirely determined by the aforementioned two parameters. One benefit of the simplified rainflow counting approach, in contrast to alternative counting methods, is that the calculation results encompass complete cycles. Following the counting typical segments, the subsequent repetitions only need to consider the number of repetitions.

Table 4.1 Rainflow counting results

Cycle	Stress range	Mean value
ADA′	9	0.5
BCB′	4	1
EHE′	7	0.5
FGF′	3	0.5
IJI′	2	– 1

4.3 Basic Protection Design

4.3.1 Basic Anti-collision Facilities

1. **Overview of anti-collision design**

The inevitable proximity and interweaving of offshore wind farms with waterways and fishing grounds provide a risk of collision between wind turbines and associated vessels. When a vessel beyond the collision prevention standards strikes the foundation of a wind turbine, it will cause significant damage to the foundation of the wind turbine. Consequently, it is imperative to implement anti-collision designs for offshore wind turbines to guarantee that the anti-collision facilities can provide good protection for the foundation structure and ships during the construction and use periods of the wind turbines, thereby minimizing damage to facilities and facilitating repairs.

In evaluating the consequences of ship collisions, the influence of floating debris on offshore wind turbines should also be taken into account. Special designs are required for the foundation structure in frozen regions to mitigate the effects of sea ice.

The wind farm exhibits an extensive range, with a separation of 500 to 1000 m between each foundation. The wind turbines at the periphery of the wind farm are the most susceptible to impact. If each foundation is engineered to a high quality for anti-collision, the project costs will be significantly elevated; hence, in most instances, anti-collision design is often implemented primarily for non-navigable openings akin to bridges.

2. **Wind turbine foundation anti-collision design standards**

The determination of anti-collision design guidelines must take into account both the probability of collision and the force of ship impact.

The occurrence of wind turbine foundations with ships is an unpredictable event, and a risk assessment can be performed for it. Evaluate the anticipated damage to wind turbine foundations resulting from ship impact according to certain impact criteria, compare it with the corresponding protection cost analysis, and obtain a rational balance point to serve as the design anti-collision standard. Once the anti-collision standards are established, the impact force of the wind turbine foundation can be computed accordingly, enabling the subsequent design of the wind turbine foundation's anti-collision features.

(1) Probability analysis of ship collision with wind turbine foundation

The probability of ship yaw towards the wind farm side is denoted as P_{Ai}. To ascertain the probability of ship deviation, the subsequent facts need to be gathered:

- The situation of shipping accidents near the wind farm, including ship collisions, ship collisions with navigation marks, ship collisions with lighthouses, and ship collisions with other offshore buildings;

- Navigation survey, investigating the scale, tonnage, frequency, quantity, and level of ship traffic management of various routes and navigation vessels in the nearby waterways;
- The navigation planning of nearby waterways should also consider future navigation planning. The probability of ship yaw can be defined as the ratio of the number of times a ship has deviated to its passing frequency. The probability of ship yaw only needs to be studied when the wind field is located in the channel and flows downstream (the ocean current flows towards the wind farm).

The probability of ships entering the wind farm at a safe distance, attributable to the failure of various warning measures within the wind farm, is denoted as P_{Ii}. When the warning measures of the wind farm fail or the ships fail to respond effectively to the warning, the ships that have strayed will approach the wind farm from a safe distance. The probability must be assessed by evaluating the reliability and efficacy of wind farm warning measures.

The probability of a ship entering a wind farm and colliding with the foundation of a wind turbine is denoted as P_{Ci}. This probability pertains the probability that a ship enters the wind farm and its trajectory passes through the foundation of a wind turbine, representing a geometric probability and depends on the layout of the wind farm and the type of ship. When a ship enters the wind farm from the boundary and adheres to a uniform distribution, the ratio of the total area of the ship's floating surface passing through the wind turbine foundation to the total area of the wind farm can be utilized as the probability of the ship entering the wind farm and colliding with the wind turbine foundation.

The aforementioned events are mutually independent, and the product of the probabilities of these events can be used to calculate the probability P of a specific type of vessel colliding with the foundation of a wind turbine, given by:

$$P = P_{Ai} P_{Ii} P_{Ci} \tag{4.40}$$

A certain safe distance exists between the waterway and the wind farm, which is outfitted with a warning system. Once the controllable state of the ship deviates, it can realign with the route through prompt warning. A ship experiencing loss of control is in a floating state after yawing, and the water flow velocity may be considered as the ship velocity for calculating the impact force. Utilizing a specific ship impact force model (relevant standard methods) or finite element numerical analysis, one can ascertain the ship impact force and its associated probability in relation to a defined ship impact standard.

(2) Calculation method for ship impact force

Three methodologies exist for calculating ship impact force: momentum formula (as utilized in highway bridge regulations), kinetic energy formula (as applied in railway bridge regulations), and vibration formula. Drawing on the expertise of port workers, this article lists the kinetic energy theory, which accounts for the energy distribution of anti-collision devices and supporting structures, serving as a reference for determining the impact force of wind turbine foundation ships.

The effective impact energy when a ship dock is:

$$E_0 = \frac{\rho}{2} M v_n^2 \tag{4.41}$$

Fenders and supporting structures will undergo deformation as a result of ship hits, with the deformation energy equating to the effective impact energy.

$$H = k_1 y_1 = k_2 y_2 \tag{4.42}$$

$$\frac{1}{2} k_1 y_1^2 + \frac{1}{2} k_2 y_2^2 = E_0 \tag{4.43}$$

In the formula, k_1 and k_2 are the elastic coefficients (KN/m) of the supporting structure and fender; y_1 and y_2 are the deformations of the supporting structure and curved structure (m); E_0 is the energy absorbed by the supporting structure and fenders after the deformation of the ship (KN m); H is the impact force of the ship (KN); ρ is the effective kinetic energy coefficient, taken as 0.7–0.8; M is the weight of the vessel (t), calculated based on the full load displacement; v_n is the normal velocity (m/s) at the time of ship collision.

During the design phase, the ship's tonnage and collision speed can be determined based on the defense target's vessel standards. First, an initial assumption is made regarding the elastic modulus of the pile. Subsequently, using the elastic modulus of the anti-collision equipment (typically provided by the manufacturer of the equipment), the deformation of the supporting structure can be calculated. This allows for the determination of the elastic modulus of the supporting structure based on the calculated displacement. The newly calculated elastic modulus is then compared with the initially assumed value, and iterative calculations are performed until the two values converge within an acceptable error range. The elastic modulus and deformation obtained from the iteration can be used to calculate the forces acting on the supporting structure.

3. **Common anti-collision facility designs**

Anti-collision facilities can be categorized into separate and attached types based on their association with the infrastructure.

(1) Separate protection system

- System mooring protection system: composed of floating body, steel wire rope, and anchoring object. Floating body movement, deformation of steel wire rope and the movement of anchored objects under collision forces can absorb a large amount of energy, which also provides good protection for colliding ships. This system occupies a large water area and is complex to construct, generally only suitable for larger ships with a bulbous head.

- Group pile pier protection system: Independent steel pipe pile foundation anti-collision piers are used, and the foundation piles consist of inclined piles that bear pressure and vertical piles that bear tension. The group pile pier structure has high stiffness, and once a collision accident occurs, the damage to the ship is relatively large. Therefore, this protective system is only suitable for situations where the probability of collision is low and other protective measures are not economical to use.
- Single row pile protection system: steel pipe piles arranged at intervals serve as anti-collision facilities. Steel pipe piles are linked by anchor chains or horizontal steel pipes, and the assessment of anti-collision capacity does not take into account the stiffness between piles, that is, the anti-collision capacity is determined based on each individual pile. The single row pile protection system is capable of withstanding the impact from small ships, and only serves as a warning and buffering function for medium and large ships.

(2) Adhesive protection system

When the impact energy is minimal, the resistance capacity of the foundation structure is substantial, or due to geological conditions, it is not easy to implement a separate independent collision avoidance system, an attached protection system can be used, which uses the foundation structure itself as a supporting structure without the necessity for separate foundation remediation efforts.

The primary focus of designing an adhesive protection system is a buffer device. The buffering device mainly utilizes two forms: a steel casing and the installation of anti-collision rubber fenders, which effectively safeguard both the foundation structure and the ship, which has been widely used in bridge engineering.

The aforementioned structures can also be divided into two categories in terms of their ability to withstand ship collision energy: one is fully capable of resisting ship impact energy, including artificial island protection systems, thin shell sand cofferdam protection systems, floating body mooring protection systems, group pile pier protection systems, etc.; Another type as a structural energy dissipation part to absorb impact energy and mitigate structural impact force, such as single row pile protection systems and attached protection systems.

4.3.2 Anti-erosion and Protective Measures

Upon the completion of the foundation construction for offshore wind turbines, the movement of water particles induced by currents and waves will be significantly influenced due to disturbances and permanent obstacles to the surface soil. Initially, a horseshoe shaped vortex will develop in front of the foundation; Subsequently, vortices (Karman vortex streets) will emerge at the base of the backflow; finally, the streamlines on either side of the foundation will contract. The alternation in local flow state will elevate the shear stress of the water flow on the bottom bed, leading to an increase in the sediment carrying capacity of the water flow. If the bottom bed is susceptible to erosion, local erosion pits will develop on the foundation, thus compromising its stability.

According to the potential flow theory, the local velocity induced by the flow around a cylinder will increase by nearly twice, resulting in an overall local velocity will be amplified to twice the original velocity. According to the principle of sediment carrying capacity, the sediment carrying capacity is directly proportional to the third power of the flow velocity. Consequently, it may be inferred that theoretically, the local sediment carrying capacity of the analysis foundation is approximately eightfold that of the original.

The DNV (Det Norske Veritas) formula, widely utilized in offshore wind power in Europe, will be listed below. The empirical formula recommended by DNV offshore wind power design specifications (under wave/current action) is

$$\frac{S}{D} = 1.3\{1 - \exp[-0.03(KC - 6)]\}, KC \geq 6 \tag{4.44}$$

In the formula, S is the scour depth (m); D is the diameter of the vertical pier column (m).

$$KC = \frac{u_{\max} T}{D} \tag{4.45}$$

In the formula, $u_{\max}$ is the maximum velocity of the water flow near the bottom of the seabed (m/s); T is the period of waves/currents.

The radius of the scour hole can be estimated as:

$$\gamma = \frac{D}{2} + \frac{\mathrm{S}}{\tan \varphi} \tag{4.46}$$

In the formula, φ is the angle of repose of sediment at the bottom of the seabed.

According to relevant references, a basic estimation indicates that $1.3D \sim 2.0D$, which determines the maximum scour depth. Given the scarcity of existing offshore wind farms in China and the limited engineering expertise for reference, as well as the uncertainty of marine hydrodynamic environments across regions, it is recommended to perform physical and mechanical model tests or three-dimensional numerical simulations of tidal current, sediment, and erosion and deposition analysis to obtain relatively accurate erosion conditions. The structure, seabed, and marine hydrological characteristics of the proposed project should be simulated through water tank tests based on relevant proportional relationships to obtain local erosion conditions. After the construction of the wind farm, regular inspections and measurements should be conducted to understand the erosion situation around the wind turbine foundation, and records should be maintained to enable feedback and rectification of theoretical outcomes.

1. **Stone throwing and erosion protection**

A straightforward technique to mitigate local erosion surrounding the pile is to reinforce it by dispersing stones within a certain range around the pile. Currently, there are relatively few effective ways for assessing the erosion prevention of stone throwing. This book primarily draws on the calculation formulas of slope protection and toe protection block stones in seawall design influenced by water flow to estimate the particle size of stone throwing:

$$d = \frac{V^2}{C^2 2g \frac{\gamma_s - \gamma}{\gamma}} \tag{4.47}$$

In the formula, *d* is the converted diameter, and for irregularly shaped stones, they are calculated as spherical shapes (m); *V* is the flow velocity of the water around the pile, and it is recommended to consider a twofold increase in local flow velocity caused by the pile foundation (m/s); γ_s is the weight of the thrown stone (kN/m^3); γ is the weight of water (kN/m^3), and seawater can be calculated at 10kN/m^3; *C* is the stability coefficient of rock throwing motion, usually taken as 0.9; *g* is the acceleration due to gravity, taken as 9.81 m/s^2.

The soil condition around the pile can be enhanced following rock throwing. The theoretical calculation of the erosion range should take into account the soil improvement caused by rock throwing, and the protection range of rock throwing may be appropriately diminished relative to the theoretical calculation range.

2. **Protection of geotextile bag filling material**

Filling geotextile bags (such as woven geotextile bags and geomembrane bags) with different materials like concrete blocks, stones, sand, and soil has a long history of application for erosion control in wharfs, embankments, or nearshore projects. These bags offer good integrity, are easy to construct, have great flexibility, and possess strong adaptability to deformation.

When using geotextile bags for protection, the scope of protection should be determined through an investigation of local erosion surrounding the foundation, and the anti-floating, anti-erosion, and anti-lifting stability of individual geotextile bags should be calculated. It can typically be designed as 0.6–2.0m for a single geotextile bag can be used for filling, or it can be designed as a large volume geotextile bag filled with filling material for filling.

For individual geotextile bag filling, relevant process tests should be performed on the drop distance of the throwing body caused by water flow prior to construction. In several ongoing projects, empirical formulas for calculating drop distance have been derived from the experimental data pertaining to throwing and filling technology, mainly associated with water flow velocity, water depth, throwing and filling monomer mass, density, and other relevant factors.

The choice of geotextile bag filling materials for offshore wind turbine foundations must consider cost-effectiveness, availability of local materials, and ease of construction. In different areas within the protection range, different filling materials can be considered. Near the pile perimeter area, the geotextile bag unit can be larger, and the filling material density should also be higher; In areas far from the pile perimeter, the individual geotextile bag can be smaller, but it should meet stability requirements. The infill material should be predominantly sand and soil as much as possible to ensure a smooth transition between the geotextile bag and the bed. For example, the protection range of the main tower pier of Strong Bridge is divided into core area, permanent protection area, and apron area, with geotextile bag sizes of 20m bagged sand, 10m bagged sand, etc.

Regardless of size of the geotextile bags, as well as geotextile woven bags or geomembrane bags, the performance requirements of geotextile bags themselves are relatively stringent. Currently, there are no mature regulations in China for anti-erosion protective geotextile bags. Consequently, based on research on the performance of more mature geotextile bags, the selection of geotextile bags involves an examination of their tensile strength, joint tensile strength, elongation, permeability, CBR (bearing ratio) bursting strength, dynamic cone failure test, wear resistance, UV resistance (which may not be considered for offshore wind turbine foundations), etc. should be investigated.

3. **Reserve flushing depth**

In the field of ocean engineering, structural design typically involves reserving a specific scouring depth to address scouring issues structurally. The primary reason is that marine engineering pile foundations typically extend deeper into the soil, resulting in a foundation stiffness that is comparatively less influenced by erosion, thereby compensating for the erosion's impact by augmenting the steel content.

In the case of pile foundation such as tripod foundations and high pile cap foundations, etc., the spacing between individual pile foundations is considerable, resulting in minimal mutual interaction. Only the local erosion of an individual pile foundation requires estimation during calculations. Given its comparatively small diameter and restricted erosion depth, reserving erosion depth is the optimal strategy.

When designing, the maximum scour depth can be determined through calculations or physical model experiments, or a certain depth may be regarded as a safety margin. The comprehensive modelling calculation can be executed in accordance with the design maximum scour line to verify that the strength, deformation, stability, frequency, and other parameters satisfy the requisite standards.

In addition to the common anti erosion measures mentioned above, there are also underwater concrete bottom protection, precast concrete blocks (similar to stone throwing mechanism), steel cages or geogrid cages for stone loading, protection, and some research units are researching and piloting biomimetic systems for anti-erosion, which reduce water flow velocity and promote sedimentation by setting up biomimetic aquatic plants, algae, etc.

4.4 Summary of This Chapter

This chapter introduces the foundation structure of offshore wind turbines, delineates their respective characteristics, proposes dynamic design numerical solutions and fatigue design methods for foundation design, and elaborates on foundation anti-collision measures and anti-erosion design for foundation protection design.

References

1. Hu HY. Fundamentals of mechanical vibration. Beijing: Beihang University Press; 2005.
2. Chen CY. Fatigue and fracture, 1st edn. Wuhan: Huazhong University of Science and Technology Press; 2001.

Chapter 5
Construction of Offshore Wind Turbine Foundation Structure

5.1 Construction of Offshore Wind Turbine Foundation Structures

The intricate and ever-changing environment, geological conditions, wind turbine capacities, and resources for construction and installation at offshore wind farms necessitate a variety of foundation structures for wind turbines. Currently, the construction of offshore wind turbine foundations is primarily classified into several types: single pile foundations, multi-pile foundations, multi-pile cap foundations [1–3], jacket foundations, gravity foundations, floating foundations, and bucket foundations, as outlined by DNV standards.

Table 5.1 provides recommended water depths for each foundation type. The structural properties of these foundations vary significantly, and the construction technology plan will differ based on the applicable water depth. This includes considerations for the foundation structure, construction equipment, and operating environment. This chapter will focus on introducing the principal construction plans for wind turbine foundations, both domestically and internationally, through illustrative examples.

According to the economic analysis results of offshore wind farm construction costs by DNV-GL, the foundation construction costs of offshore wind farms account for approximately 6% of the total investment cost of the wind farm. There are significant differences in the construction costs of various foundation structures, so it is essential to fully consider the construction processes and complexities when designing the foundations. Determining the most suitable foundation structure for the wind farm conditions and how to leverage existing construction equipment to achieve the lowest foundation construction costs and optimal efficiency has long been one of the goals of discussion and effort within the industry.

M. Ge et al., *Offshore Wind Turbine Technology*,
https://doi.org/10.1007/978-981-96-2175-0_5

Table 5.1 Recommended water depths for different types of foundations by DNV standard specifications

Infrastructure	Water depth (m)
Gravity based foundation	0–10
Bucket shaped single column foundation	0–25
Single column structure (single pile, three pile) foundation	0–30
Three legged/four legged conduit rack foundation	> 20
Floating foundation	> 50

5.1.1 Construction of a Single Pile Foundation

A single pile foundation, also known as a single steel pipe pile foundation, is characterized by its lightweight construction, straightforward design, and well-defined stress distribution. The pile diameter of a single pile foundation is substantial, with thick walls, and it can be either driven into the seabed using a pile hammer or drilled into the seabed, or a combination of both methods may be employed. In offshore wind power projects, the pile diameter typically ranges from 4 to 6 m, with a maximum diameter of approximately 8 m. This diameter is primarily constrained by the capacity of construction equipment, such as pile hammers and cranes.

Currently, hydraulic hammers are predominantly used for pile driving in the construction of single pile foundations for offshore wind farms in China. The largest hydraulic hammers available include the IHCS2000 and Menck S1900, among others. The diameter of the top flange of the installed foundation can reach up to 5.5 m. For driving larger diameter foundation piles, it is essential to create a matching replacement ring sleeve on the upper flange of the foundation.

The typical depth of mud penetration for a single pile foundation range from 18 to 50 m. The thickness and depth of immersion of an individual steel pipe pile depend on several factors, including the foundation design load, soil conditions at the site, water depth, surrounding environment, and applicable design standards. Figure 5.1 illustrates the offshore wind turbine featuring a single pile foundation.

To assure the safe and steady functioning of wind turbines, the levelness criteria for single-pile foundations are severe, with the prevalent industry standard requiring that the levelness should not exceed 2 ‰. Transition parts, mainly used in offshore wind farms abroad, priminarily serve to modify the top flange linked to the tower to obtain the needed levelness. In the offshore wind farms already erected in China, a revolutionary single pile foundation devoid of a transition section has been successfully developed and extensively applied, with a levelness of up to 0.2 ‰, much below the levelness requirement of the foundation top flange. This type of foundation structure removes the high-strength grouting process in the transition section, considerably lowering the installation time and cost of the foundation structure, and may achieve a foundation installation efficiency of 1 day per wind turbine. However, the use of this infrastructure is presently mostly concentrated on intertidal zones,

Fig. 5.1 Offshore wind turbine with single pile foundation

and the outcomes of future applications in nearshore and more severe wind and wave conditions still need to be tested in batches.

According to industry experience, single pile foundations are frequently utilized in offshore wind farms at water depths of up to 20 m, with examples in Europe, where they are employed to depths of 40 m. Due to the fact that the piles and towers of the single pile foundation are both tubular, it is very convenient to connect them on site by directly employing flange bolts. To assure the levelness of the infrastructure, transition sections are generally deployed internationally. Transition segments are rarely utilized domestically to maintain stability.

Explore the level of infrastructure and build some process equipment that is suited for the national conditions. Figure 5.2 displays an single pile foundation of offshore wind turbine in China.

1. **Construction process of single pile foundation**

The main construction procedure of a single pile foundation, namely one with transition section will be further upon in detail below.

(1) The transportation and movement of single pile foundations on the dock

The volume and mass of a single pile foundation are relatively large, requiring the use of heavy transportation equipment to move from the dock storage area to the designated installation location. As shown in Fig. 5.3, when moving the single pile at the dock, heavy flatbed trailers or large gantry cranes are generally used.

(2) Loading at the dock

According to the conditions of the dock equipment, multiple options exist for single pile foundation dock loading.

Fig. 5.2 Single pile foundation of a China offshore wind farm

Fig. 5.3 Land transportation of single pile foundation

① The use of onshore heavy-duty cranes for foundation loading, as shown in Fig. 5.4, is often used when transporting large transport barges without cranes. The following points should be noted:

Fig. 5.4 Heavy duty crane on shore for foundation loading

- The operational range of the dock crane;
- The load-bearing capacity around the dock where the crane is located;
- Pay attention to the collision risk between single piles and ships, as well as the upper structure of ships during the lifting process;
- Select the appropriate lifting capacity of the crane based on the height of the surrounding structure of the ship's hull;
- Pay attention to the rising and sinking movements of ships due to changes in tide levels.

② When transferring single piles on large transport ships or self-raising installation ships with cranes, use the massive crane aboard the installation ship to load the ship. Currently, the presence of cranes resulted in a lower deck area on the ship compared to specialized large barges, thereby constraining the number of single piles that can be loaded is reduced. This technique demands drastically decreased requirements for dock cranes and dock load-bearing capability.

③ Using self-contained modular transporters:

- It can be used for loading large barge transportation methods;
- It can directly transport from the yard to the barge deck;

- A bridge deck needs to be constructed between the dock and the barge for transportation;
- It's very sensitive to changes in tide levels, suitable for loading at inland ports;
- During the loading process, safety must be strictly controlled.

(3) Single pile sea transportation

The singular pile sea transportation techniques can be divided into three types: transport ship transportation, installation ship transportation (refer to Fig. 5.5), and floating towing method (refer to Fig. 5.6).

(4) Foundation turning and pile sinking

When flipping the single pile foundation, two cranes aboard the ship are employed to lift the upper and lower ends of the single pile separately.

The crane hauling the single pile's upper end should rise gently first. After the single pile is hoisted about 1 m off the deck, the crane clutching the lower end of the single pile should stop ascending. The crane clutching the upper end of the single pile will continue to rise slowly. As it approaches vertical alignment, the crane at the lower end of the single pile will pull away, transitioning to a complete lifting of the single pile by the crane above. The crane will then carefully move the single pile towards the designated pile driving point. Once there, the pile will be seized by the pile grabber, employing the weight of the single pile to sink down. The vertical alignment will be confirmed using sensors and detecting equipment on the pile grabber, as shown in Fig. 5.7.

The primary method for arranging a single pile for turning over involves use a crane to elevate the lifting ears welded on the outer wall of the single pile for turning over and move, as shown in Fig. 5.8.

Fig. 5.5 Installation ship transportation

Fig. 5.6 Floating and dragging method

Fig. 5.7 Pile holder

When employing bucket wall lifting ears for hoisting, due to the large foundation mass, it is essential to weld larger lifting ears onto the bucket wall. This weld seam may affect the foundation structure during pile driving. Therefore, many projects use professional single pile foundation turning lifting equipment (see Fig. 5.9) or

Fig. 5.8 Single pile flip positioning

specialized single pile foundation turning brackets on board (see Fig. 5.10) to turn and position single pile foundations for now.

(5) Pile driving operation

Upon completing the foundation positioning and pile sinking, retain the single pile foundation at the pile driver location. Once the crane unloads the lifting device/beam, lift the pile hammer and place it on the top of the single pile for pile driving. Following multiple pile driving cycles, it is essential to pause and calibrate the verticality until the individual pile reaches the required depth.

From the standpoint of construction methods, the present construction plans for single pile foundations are largely separated into two types: hydraulic hammer (vibratory hammer) pile drive and drilled pile construction.

Hydraulic hammer pile drive is the most extensively used technology, noted for its great construction efficiency and inexpensive cost, with reduced technical requirements compared to drilled pile construction. However, its downsides include building noise, which may impact adjacent marine life. Therefore, a thorough environmental evaluation must be undertaken before construction, and appropriate sound-dampening equipment should be employed to reduce noise if necessary. For instance, a certain project in a coastal area was postponed since the pile driving procedure could adversely damage native white dolphins.

Fig. 5.9 Professional single pile foundation turning lifting device

Fig. 5.10 Ship specific single pile foundation turning bracket

After transporting, turning over, and sinking the single pile to the machine position, use a pile hammer to construct the single pile. Generally a hydraulic hammer or vibrating hammer is used. The hydraulic hammer and vibrating hammer are shown in Figs. 5.11 and 5.12, respectively.

Fig. 5.11 Hydraulic hammer

Fig. 5.12 Vibration hammer

The process of employing a drilling machine to drill holes on the seabed, installing individual piles, and subsequently pouring concrete is referred to as drilling and pile driving. This approach is commonly used for more challenging seabed.

(6) Transition section installation

After detaching the pile hammer from the crane, the transition section is progressively lifted and brought towards the placed single pile. The transition piece is then fitted over the single pile and lowered to the specified position. After validating the vertical alignment, it is held in place while high-strength concrete is injected using grouting equipment to connect the transition portion with the single pile. The crane can only be released when the high-strength concrete has cured for roughly one week and has reached the requisite strength.

(7) Stone throwing protection

In single pile foundations, an increase in pile diameter heightens the risk of scouring the seabed around the pile. Therefore, upon the completion of foundation installation is completed, a specialized stone throwing ship must be used to safeguard the vicinity of the single pile foundation. The distance of stone throwing is contingent upon the pile diameter and seabed conditions. Due to the fact that submarine cables need to enter wind turbines through submarine cable ducts at the bottom of the foundation, stone throwing protection should generally be implemented following the completion of submarine cable connection construction.

2. **Advantages and disadvantages of single pile foundation construction**

(1) Advantages

- There is basically no need for construction preparation and no need to tidy up the seabed.
- Low-manufacturing cost, convenient transportation, simple construction process, and most widely used in shallow water (water depth less than 20 m) development.

(2) Disadvantages

- Due to its structural characteristics, anti-erosion protection is required after installation.
- Due to limitations in construction equipment, it is not suitable for large offshore wind turbines (5 MW/6 MW level).
- Not suitable for construction in areas with large rocks on the seabed.

Single pile foundations are currently the most widely used foundation structure in offshore wind farm projects both domestically and internationally. The 2.5 MW wind turbine intertidal zone project by Goldwind Technology Co., Ltd. utilizes single pile foundations.

5.1.2 Construction of Multi-pile Foundations

Multi-pile foundations refer to a form of offshore wind turbine foundation structure consists of three or more single pile foundations. Common layouts include three-pile, five-pile, or six-pile foundations, with three-pile foundations being the most popular in present wind farms (see Fig. 5.13). In a three-pile foundation, the center single pile is utilized to support the wind turbine, similar to the single pile foundation structure, while the surrounding three piles provide support for the central pile. The single pile and three-pile foundation are joined utilizing high-strength grouting materials. Multi-pile foundations are developed based on single pile foundations, boosting the stiffness and strength of the surrounding piles. They are suited for water depths equivalent to those of single pile foundations, however because multi-pile foundations use smaller diameter piles, the requirements for pile driving equipment and construction are lower compared to single pile foundations.

Fundamental Principle of Multi-Pile Foundation Design: Multiple medium-diameter steel pipe piles are installed on the seabed, typically buried 10–20 m below the surface. These steel pipe piles are evenly arranged, with their tops connected to a multi-pile foundation sleeve using high-strength grouting material, thereby forming a composite foundation. The multi-pile foundation structure consists of prefabricated welded components. Due to the numerous connection points and complex shapes involved, manual welding is often required, which can lead to increased manufacturing costs and operational challenges. The main framework of the central foundation is connected to the upper tower to bear the load, which is then distributed to the surrounding steel pipe piles. The diameter of the steel pipe piles generally ranges from 2 to 3 m.

Fig. 5.13 Three pile foundation offshore wind turbine

The installation and construction methods for multi-pile foundations can be primarily categorized into two main types.

(1) Pile first, then install the foundation

- Complete the small pile driving operation for multi pile foundations, with the same driving process as single pile foundations, but due to the pile diameter is significantly lower than that of a single pile, resulting in reduced difficulties and equipment needs for pile driving, hence facilitating the process. When adopting this approach, the placing of small piles is often performed by a temporary steel positioning frame. Prior to driving each tiny pile, install the positioning frame on the seabed to achieve the positioning of each small pile, and subsequently drive each small pile into the its allotted hole independently.
- Lift several pile foundations and slowly install them into the pile group by positioning piles.
- When the pile sleeve (pile shoe) enters the already driven pile and reaches the limit, the foundation is fixed and sometimes numerous pile foundations collapse directly onto the seafloor, which bears the major gravity.
- Fully weld and seal the top and lower ends of the pile sleeve, pour high-strength concrete, and after a time of cooling and solidification, finish the connection between the multi pile foundation and the pile sleeve. During this phase, the crane remains motionless until completion, unless the multi pile foundation directly falls on the seafloor, which bears the major gravity.

(2) First lift the foundation, then drive the pile

- The multi-pile foundation is positioned at the designed location by first hoisting it into place, allowing it to rest directly on the seabed, which supports its weight.
- The group piles are then driven into the pile sleeve through precise positioning, using a piling process similar to that of single pile foundations.
- Once the group piles reach the designed depth, the upper and lower ends of the pile sleeve are fully welded shut, and high-strength concrete is poured in. Temporary or permanent stabilizers are generally used on the upper part of the pile sleeve to ensure that it remains centered during the grouting process. After a period of time, once the high-strength concrete has cooled and solidified, the connection between the multi-pile foundation and the pile sleeve is complete.

Multi-pile foundations are utilized in Europe as a transitional form from single pile foundations to jacket foundations. However, because to the absence of economic advantages in manufacturing, they are currently rarely utilized in Europe. In China, due to the unique geological conditions in the Jiangsu region, multi-pile foundations have been widely applied in high-capacity offshore wind turbine projects, such as the 4 MW, 5 MW, and 6 MW wind turbines in the Jiangdong intertidal demonstration wind farm, as well as the 2.5 MW test wind turbines from Goldwind Technology Co., Ltd. In Germany, the first offshore wind farm, Alpha Ventus, featured tripod

Fig. 5.14 German offshore wind power plants

foundations for six of its original offshore wind turbines (by Multibrid), as seen in Fig. 5.14.

Multi pile foundation has the following advantages:

- The total structure exhibits commendable stability and considerable bending resistance in comparison to single pile foundations;
- The foundation design of offshore wind turbines is informed by the gained from offshore oil platform design;
- Suitable for water depths of 15–30 m, the control of foundation levelness is achieved using fixed offshore platforms including floating docks;
- Due to the fact that the depth of steel column embedding is related to the geological characteristics of the seabed, tripod structure foundations are typically unsuitable for sites with extensive areas of rock on the seabed.

5.1.3 Construction of Multi-pile Foundation

A multi-pile foundation, or group pile high pile foundation, is frequently used in offshore bridge support structures. The specifications for multi-pile cap foundation construction are presented in Table 5.2.

The offshore wind farm of Donghai Bridge is shown in Fig. 5.15.

Table 5.2 Construction requirements for multi pile foundation

Construction description	Construction requirements
Basic construction process	(1) Installation of pile foundation (pile driving) (2) Place steel casing and pour concrete cushion layer (3) A protective layer against erosion composed of stone throwing

Fig. 5.15 Donghai Bridge offshore wind farm

The foundation of group pile high pile cap typically adopts multi steel pipe piles, and the cap is made of reinforced concrete structure. To counteract lateral forces, the steel pipe piles need to be inclined at a certain angle. Upon completion of the foundation construction, a certain thickness of anti-erosion material needs to be laid on the seabed.

1. **Construction technology of multi pile cap foundation**

- Sinking, positioning, and driving the pile. The pile driving construction process is roughly the same as the single pile foundation construction process.
- After completing the pile group, first cut the surface of the pile group flat, build a steel cage at the top of the pile group, and wrap the pile head inside.
- Wrap the constructed steel cage with a steel casing, install a sealing plate, and pour concrete (see Fig. 5.16).
- Upon achieving the design strength of the concrete and completing the pile sinking, the steel casing procedure is employed to install the sealing plate and pour

Fig. 5.16 Concrete pouring for multi pile foundation

the bottom sealing concrete; Once the strength of the bottom concrete reaches the design strength, seal the water hole, use the air lift method to remove the sediment inside the pile, and pour the pile core concrete. The upper 2–3 m of the pile core needs to be compacted using vibration.

- Place the support base of the transition section of the wind turbine tower on the bottom concrete, weld and fix the support base and steel pipe piles, and then lift the transition section of the tower.
- After installing the transition section, tie the steel bars of the bearing platform in one go and pour the concrete of the bearing platform again.
- After the concrete strength of the bearing platform meets the design requirements and is cured, the steel pipe box will be entirely dismantled. The steel pipe box is reusable.

2. **Advantages and disadvantages of multi pile foundation**

(1) Advantages

- Steel pipe piles are generally angled to resist lateral forces.
- The diameter of steel pipe piles is small, facilitating production and transportation.
- The complexity of construction technology is less than that of single pile foundation.
- The construction process is relatively mature.

(2) Disadvantages

- The pile foundation is relatively long, and the overall structure tends to be heavy and heavy.
- Suitable for water depths ranging from 5 to 20 m.
- Due to the wave induced pushing force on the pier, the connection between the pile and the pier needs to be reinforced.

5.1.4 *Construction of Jacket Foundation*

The jacket foundation consists of an open steel truss welded from a series of tubular components that can reach from the seabed to the water surface, with piles driven into the seabed from each leg of the jacket. The offshore wind turbines at the CGN Yangjiang Nanpeng Island wind farm use this type of foundation, as shown in Fig. 5.17.

1. **Installation and construction method of jacket foundation**

There are two types of installation methods for jacket foundation.

(1) The steps for piling first and then lifting the foundation are as follows:

- Drive a few small piles, using the identical piling process as a single pile foundation;
- Lift the jacket and gradually install it onto the pile group by positioning piles;
- When the pile sleeve (pile shoe) enters the previously driven pile and reaches the limit position, it remains fixed;
- Completely weld and seal the upper and lower ends of the pile sleeve, and pour high-strength concrete within it;
- After a period of time, it can cool and solidify, during which the crane cannot be released.

Fig. 5.17 Foundation of CGN Yangjiang Pengdao offshore wind turbine

(2) The steps for hoisting the foundation first and then piling are as follows:

- First lift the conduit rack to the design position;
- By positioning, the group piles are driven into the pile sleeve, and the driving process is the same as that of a singular pile foundation;
- Upon attaining a specified position, thoroughly weld and seal the upper and lower ends of the pile sleeve, and pour high-strength concrete into it;
- After a while, it can cool and solidify, during which the crane cannot be released.

2. **Advantages, disadvantages, and scope of application of jacket foundation construction plan**

(1) Advantages

- Employing small-diameter steel pipes for insertion, with terminal filling or forming connections, suitable for deeper water regions.
- The bearing capacity of the covering layer is substantial, and the requirements for pile driving equipment are comparatively minimal.
- The conduit rack is processed in the factory and transported and installed as a whole.

(2) Disadvantages

- The on-site operation time is relatively long.
- The foundation structure of the jacket rack is large and heavy.
- Expensive equipment is required for transportation and lifting.

(3) Scope of application

So far, the optimum depth range for jacket foundations is roughly 50 m. The two offshore wind farms that have utilized jacket foundations in the deepest waters are Beatrice (with a water depth of 45 m) and Alpha Ventus (with a water depth of 30 m), both sustaining 5 MW wind turbines. Jacket foundations are also often utilized to support offshore substations. Although financial concerns limit the development of jacket foundations in waters less than 100 m deep, they can nevertheless be employed in deep water environments (above 100 m).

Using vessels: barge + crane or floating crane.

3. **Typical application cases**

The overall offshore installation case of China Communications Construction Offshore Wind Power Construction Technology Research and Development Center Daishan Base is shown in Fig. 5.18. Two 5 MW offshore wind turbines have been installed, as demonstrated. The wind turbine adopts a jacket foundation and is installed with a floating crane that has a lifting capacity of 4000 tons.

Fig. 5.18 Case study of offshore integrated installation at Daishan Base of China Communications Construction Offshore Wind Power Construction Technology R&D Center

5.1.5 Gravity Based Foundation Construction

Gravity foundations (see Fig. 5.19) utilize a reinforced concrete construction, relying on their own weight to produce stability and resist the combined loads of wind and waves on wind turbines. Early offshore wind farms in the world largely used gravity foundations, which are composed of reinforced concrete and are suited for shallow waters, ranging from 0 to 10 m. The conveyance of gravity foundations is represented in Fig. 5.20. The floating conveyance of gravity foundations is demonstrated in Fig. 5.21.

1. **Construction technology of gravity foundation**

- Use an excavation ship to dig a pit on the seabed where the wind turbine is installed, with an area of approximately 50 m $\times$ 70 m and a depth of approximately 4.5 m.
- To enhance the flatness of the bottom of the pit, fill the excavated pit with crushed stones, ensuring that the plane error should not exceed 5 cm.
- Utilize a transport ship to transport the pre-assembled gravity foundation to the installation site, and lift the gravity foundation. Owing to the weight of the gravity foundation, the pit surface in contact with the foundation will sink.
- Use a pump to suck the sea sand into the hollow foundation, and once the sea sand has settled and compacted, pump out the excess water above.

Fig. 5.19 Gravity foundation

Fig. 5.20 Gravity foundation transportation

2. **Advantages, disadvantages, and scope of application of gravity-based foundations**

(1) Advantages

- No need for pile driving, directly reducing construction noise.
- Simple structure.
- The manufacturing cost is lower than that of a single pile foundation.
- Not affected by the seabed, with good stability.

Fig. 5.21 Gravity foundation floating transportation

Use of vessels: large transport ships or floating cranes.

(2) Disadvantages

- The underwater workload is substantial, necessitating dredging and stone-throwing operations on the seabed. The structural integrity and seismic resistance of the foundations are inadequate, and they are significantly affected by erosion, leading to a high consumption of various filling materials.
- Over time, gravity-based foundations experience sinking, which is influenced by their structural integrity, geological conditions, and construction techniques.
- Additionally, the transportation and installation costs associated with ships are high, time-consuming, and labor-intensive. This process requires specialized manufacturing equipment capable of supporting the weight of the foundation, such as reinforced docks or specialized transport ships for moving foundation base caissons.

Scope of application: Gravity foundations are mostly used in locations where pile driving is infeasible and in proximity to precast concrete structures nearby. Gravity foundations are frequently employed in areas susceptible to ice.

Typical case: Thornton Bank offshore wind farm is the first offshore wind farm in Belgium and the world's first commercial offshore wind farm to use gravity foundations. It is situated 27-30 km north of the Belgian coastline at a depth of 12–27 m The gravity foundation employed is a reinforced concrete structure featuring a hollow design and a transport weight of around 1200 tons. Upon installation, it is filled with fine sand or crushed stone, totaling around 6000 tons in weight of over. To install the foundation, the construction wind turbine used a total of over 100 types of vessels, including the world's largest crane ship Rambiz (with a maximum lifting capacity of approximately 3300 tons) at that time (2007).

5.1.6 Floating Foundation Construction

In 1972, Professor William E. Heronemus at the University of Massachusetts Amherst first proposed the concept of large-scale offshore floating wind turbines. After the mid-1990s, with the establishment of the commercial wind energy industry, this topic gained prominence in the mainstream research community. Global deep-sea wind energy resources are abundant, with areas at depths of up to 600 m considered the best for transmitting and generating electricity.

As water depth increases, the economic costs limit the use of steel structures. In the offshore oil and gas industry, the depth limitation for platforms is around 450 m (1500 ft, where 1 ft = 0.3048 m). In the offshore wind energy sector, depth constraints may be limited to within 100 m due to cost considerations. Floating foundations consist of floating platforms and anchoring systems. The types of floating foundation wind turbines are shown in Fig. 5.22, including cylindrical buoy structures and tension leg platforms, similar to those used in the oil and gas industry.

1. **Typical Case (I)**

Hywind is the world's first deep-water floating large-capacity wind turbine, built as an international cooperation project involving numerous countries, including Norway, Denmark, Germany, the UK, and the Netherlands, and is located in the northern waters of Norway. Statoil Hydro (Norwegian National Oil Company) implemented

(a) HECICON (b) Hywind

(c) Ideol (d) Windfloat

Fig. 5.22 Types of floating foundation wind turbines

the concept of horizontal axis wind turbines in this project. In 2009, a prototype wind turbine was placed by the Norwegian National Oil Company in the Norwegian Sea. This wind turbine has a capacity of 2.3 MW and was built by Siemens, mounted atop a floating buoy at a depth of 100 m. The buoy was made and installed by the French business Technip, while the cables were produced and placed by Nexans.

The floating foundation has a diameter of 8.3 m, with an underwater cylindrical segment spanning 100 m long, and its submerged part is mounted on a buoy over 100 m tall, anchored to the seabed with three mooring lines. The wind turbine was assembled in Stavanger, Norway, progressively constructed at sea using a crane on a ship, with adjustments made based on actual conditions. The horizontal axis wind turbine is erected in the fjord, partially immersed in seawater. Water and pebbles are used as ballast, with the wind turbine installed on top. The assembled wind turbine is transported for offshore installation, and the anchors are set in place. Hywind began functioning on September 8, 2009, generating around 9 GWh of power annually.

2. **Typical Case (II)**

WindPlus, created in 2011, is located 5 km offshore in Portugal and features the WindFloat installation, equipped with a Vestas V80 2.0 MW turbine. On October 22, 2011, Vestas announced the completion of Portugal's first offshore wind turbine. This Vestas V80-2.0 MW wind turbine is the first to be installed on the unique floating base known as WindFloat. The installation and commissioning of the wind turbine were completed on land, following which it was towed to the intended offshore site, making it the first offshore wind turbine in Portugal.

The Vestas V80-2.0 MW turbine deployed on the WindFloat delivers enough electricity to supply 1300 families and has generated over 1.7 GWh of electricity. This technology permits wind turbines to be erected at offshore places with water depths above 40 m, where stronger wind resources are available compared to shallow-water offshore wind farms.

5.1.7 Construction of Bucket-Foundation

The bucket foundation consists of a central column and a steel circular bucket. The steel circular bucket is attached to the column by vertical steel skirts and reinforcing ribs. The bucket foundation manages its level by regulating the negative pressure vacuum within each chamber of the bucket. The load from the wind turbine is transmitted through the central column to the reinforcing ribs and the steel skirts. After installation, the foundation requires friction balance between the foundation and the bucket wall.

The structural advantages of the bucket foundation include saving steel material and lowering offshore building time, which helps minimize production and installation costs. Currently, bucket foundations are mainly solely utilized to meteorological tower foundations both domestically and internationally, with no uses in offshore wind turbines.

Fig. 5.23 Bucket foundation

Construction Process: The bucket foundation consists of numerous cylindrical buckets, with the lower half of the buckets open. After towing the foundation to the appropriate place, the air inside the foundation is slowly regulated and evacuated to allow it to sink to the seabed. Air is constantly removed until the foundation is embedded in the seafloor and can give a particular bearing capacity. The bucket foundation was originally deployed to offshore oil development abroad in 1993, and there have been no applications in big wind turbine projects, with utilization limited to a few meteorological towers, as shown in Fig. 5.23.

5.1.8 Comparison of Advantages and Disadvantages of Various Basic Construction Methods

Table 5.3 presents a comparison of the advantages and disadvantages of various foundation constructions.

5.2 Installation and Construction of Offshore Wind Turbines

5.2.1 Single Blade Installation Construction

The single-blade installation refers to the separate installation of the three blades of the offshore wind turbine. The single-blade installation of the offshore wind turbine by Goldwind Technology is shown in Fig. 5.24.

Table 5.3 Comparison of advantages and disadvantages of various foundation constructions

Infrastructure	Advantage	Disadvantages
Pile	(1) No need to tidy up the seabed (2) The structure of large-diameter steel pipe pile scheme is affected by waves (3) Due to its low cost and simple process, it is currently the most widely used in shallow water (less than 20 m) development	(1) Need to prevent ocean currents from scouring the seabed (2) The application depth is less than 25 m, and due to limitations in depth and underground environmental factors, it is not widely used in deep water areas (3) Not suitable for locations with large rocks on the seabed (4) The construction difficulty is high, and the verticality control of the pile body is difficult (5) Short construction period, wind turbine installation can be started immediately after pile driving
Tripod	(1) The overall structural stability is good (with a light foundation weight) (2) The design of offshore wind turbine foundations draws on the experience of offshore oil platforms	(1) Compared to a single pile foundation, the overall quality is relatively large (2) The lifting capacity of the construction vessel is relatively low compared with that of the single pile foundation, the construction difficulty is small, and the construction period is long
Multi pile foundation platform	(1) Steel pipe columns are generally inclined at a certain angle to resist lateral forces (2) The pile foundation has a small diameter, making it easy to manufacture, transport, and lift (3) The construction process is similar to that of a single pile foundation and is relatively mature	(1) The pile foundation is relatively long, and the overall structure tends to be heavy and heavy (2) Suitable for water depths ranging from 5 to 20 m (3) Concrete pouring and curing are required, and the construction period is long after completion (4) Due to the significant pushing force generated by waves on the pier, the connection between the pile foundation and the pier needs to be reinforced

(continued)

Table 5.3 (continued)

Infrastructure	Advantage	Disadvantages
Jacket	(1) The diameter of the steel pipe is small, and the end is filled or formed for connection, suitable for deeper water areas (2) The bearing capacity of the covering layer is high, and the requirements for pile driving equipment are relatively low (3) The conduit rack is processed in the factory and transported and installed as a whole	(1) The on-site operation time is relatively long (2) The structure is huge and heavy (3) Expensive equipment is required for transportation and lifting (4) The construction period is relatively short
Gravity type	(1) No need for pile driving, directly reducing construction noise (2) Simple structure (3) The manufacturing cost is lower than the cost of a single pile foundation (4) Not affected by the seabed, good stability	(1) The underwater workload is large, the structural integrity and seismic resistance are poor, and it is greatly affected by environmental erosion. The consumption of various filling resources is high (2) As time passes, there is a problem of subsidence, which is related to its own structure, geological structure, and construction method (3) Ship transportation and foundation installation in the sea have high construction costs, are time-consuming and labor-intensive, and have high requirements for ships transporting foundation base caissons. They require specific production equipment that can withstand the weight of the foundation (docks, reinforced docks, or specialized transport ships) (4) Mainly used in sea areas where pile driving is not possible, areas with precast concrete structures nearby, and sea areas prone to icing

1. **Single blade installation method**

Two main installation methods exist for single blade blades.

- Three blades are installed in a Y-shape. First, mount the two upper blades at an angle of 30° to the horizontal line. Then, install the lower blade in a vertical orientation.

(a) (b)

(c)

Fig. 5.24 Installation of single blades for Goldwind Technology's offshore wind turbine

- All three blades are positioned horizontally. Initially, install a horizontal blade, rotate the impeller as a whole by 120°. Nest, install the second blade horizontally, rotate the impeller as a whole by 120° again, and finally, install the third blade horizontally.
- The advantages and disadvantages of the two installation methods are listed in Table 5.4.

Table 5.4 Advantages and disadvantages of two installation methods

Installation method	Advantage	Disadvantages
Y-shaped installation	No need for adaptive design of wind turbine turning gear	The installation of lifting equipment has a very complex structure and poor reliability
Horizontal installation	Fast installation speed	It is very difficult to turn a direct drive wind turbine

2. **Installation method of single blade wind turbine**
(1) The first installation method for single blade wind turbines

The first installation method of single blade wind turbine: tower + nacelle + generator + hub + blades. The installation process is as follows:

- The tower of the wind turbine is lifted separately in sections by two cranes (when transporting the tower vertically, a single crane can also be used);
- Lift the engine room, generator, and wheel hub separately;
- Lift all blades separately.

This installation method does not involve pre assembly offshore components, resulting in reduced demands on dock resources, and the quality of each component is generally minimal (for instance, a 6 MW wind turbine, where the heaviest individual component is the generator, which weighs 180 tons). The separate lifting of each component requires the development of supporting fixtures and lifting fixtures for each component, especially for the development of generator lifting fixtures and single blade lifting fixtures. Simultaneous, offshore lifting is time-consuming, exhibits low lifting efficiency, and carries a high risk of interruption in lifting work. The elevated cost renders it approximate for areas far from the coast, where transport ships can transport a maximum quantity of wind turbine components as possible at once. The installation vessel may be a self-elevating platform barge or a self-propelled installation vessel, and it is also required to design lifting points for each component of the wind turbine.

(2) The second installation method for single blade wind turbines

The alternative installation method for single blade wind turbines: tower + nacelle and generator pre-assembly + hub + blades. The nacelle and generator pre assembly is carried out onshore, and the installation process is as follows:

- The tower is lifted separately in sections using one or two cranes;
- Lifting the assembled engine room and generator;
- Lifting wheel hubs;
- Lift all blades separately.

This installation technique involves minimal pre-assembly of onshore components, hence reducing the demand for dock resources. The engine room and generator can be lifted concurrently (such as a 6 MW wind turbine with a total weight of 280 tons), and the minimum lifting capacity of the lifting crane is required to meet the lifting wheel load. The accessible lifting vessel is a self- elevating platform barge, which requires the development of assembly support fixtures for the engine room and generator, in addition to single blade lifting fixtures. The lifting efficiency is low, and

Fig. 5.25 Installation method of the third type of single blade wind turbine

the lifting expenses are elevated. Siemens implemented this installation technique in the Gun Fleet Sands project at sea.

(3) The third installation method for single blade wind turbines

The third installation method for single blade wind turbines: tower + nacelle, hub and generator pre assembly + blades, as shown in Fig. 5.25.

The procedure for installation is as follows:

- Preassembly of engine room, generator, and hub onshore;
- The tower is lifted separately in sections using one or two cranes;
- Lifting assembled engine rooms, generators, and wheel hubs;
- Lift all blades separately.

This installation method entails the pre-assembly of onshore components, and due to absence of impeller assembly, the requirements for dock resources are not high. The engine room, generator, and hub can be lifted simultaneously (such as a 6 MW wind turbine with a total weight of 370 tons). The lifting vessel that can be used is a self-elevating platform barge, which requires the development of assembly support fixtures and single blade lifting fixtures for the engine room, generator, and hub. At the same time, the wind turbine is required to have lifting points for the engine room, generator, and hub and corresponding tooling lifting equipment. Due to the need to lift each blade separately, the lifting efficiency is average and the lifting cost is average.

5.2.2 *Rabbit Ear Installation Construction*

Rabbit ear installation involves pre assembling two blades onto the wheel hub.

1. **Rabbit ear wind turbine installation method**

(1) The first installation method of rabbit ear wind turbine: tower + nacelle + generator + hub and two pre-assembled blades + blades.

Installation process:

- Preassemble the hub and two blades onshore to form a two bladed impeller (also known as a rabbit ear impeller);
- The tower is lifted separately section by section using two cranes;
- Lift the engine room and generator separately;
- Lifting two bladed impellers;
- Lift the remaining vertical blades.

This installation method comprises pre-assembly of components onshore, therefore the requirements for dock resources are often moderate. The two-blade rotor hoisting (for example, on a 6 MW wind turbine, the total weight of the three blades is 150 tons, with the generator weighing 180 tons) needs the hoisting crane to have a minimum lifting capacity of no less than 300 tons at hub height. The relevant hoisting vessels include self-elevating platform barges, self-propelled installation vessels, or crane ships. It is important to design assembly support tools for the two-blade rotor and vertical blade lifting fixtures. Additionally, the wind turbine must be supplied with lifting points and related tooling for the two-blade rotor hoisting. The hoisting efficiency is average, and the hoisting cost is similarly average.

(2) The second installation method of rabbit ear wind turbine: tower + nacelle, generator, hub and two pre-assembly of blades + blades, as shown in Fig. 5.26.

Fig. 5.26 Rabbit ear installation

Installation process:

- Pre-assembly the engine room, generator, hub, and two blades onshore to form a two bladed nose;
- The tower is lifted separately in sections using one or two cranes;
- Lifting and assembling two blade machine heads together;
- Lift the remaining vertical blades separately.

This installation process entails a large amount of onshore component pre-assembly, resulting in higher requirements for dock resources. The nacelle, generator, hub, and two blades can be hoisted together (for example, on a 6 MW wind turbine, with a total weight of 420 tons). The hoisting crane must have a minimum lifting capacity of no less than 600 tons at hub height. The relevant hoisting vessels include self-elevating platform barges, self-propelled installation vessels, or heavy lift ships.

It is important to design assembly support tools for the two-blade nacelle and vertical blade lifting fixtures. Additionally, the wind turbine must be fitted with lifting points and related tools for the two-blade nacelle assembly hoisting. The hoisting efficiency is quite high, while the hoisting cost is ordinary.

This installation technology has been utilized in numerous projects abroad, such as Horns Rev (Dong Energy) (see Fig. 5.27), North Hoyle, Barrow, Scroby Sands, and Kentish Flats, among others.

2. **Typical cases**

The Scroby Sands project is one of the earliest offshore wind farms in the UK, consisting of a total of 30 Vestas V80 wind turbines. The construction process is roughly as follows.

- Fundamental construction, employing single pile foundation. To meet the levelness requirements of the flange connecting the pile body and the wind turbine tower, a transition section is utilized. The transition section is equipped with a J-shaped cable conduit for linking wind turbines to underwater cables.

Fig. 5.27 Horns ReV wind turbine Rabbit ear installation

- Use a self-elevating barge to transport a single pile foundation weighing 200 tons to the construction site, perform pile driving operations, and install transition sections.
- Lay rocks around the single pile foundation to prevent erosion from affecting the foundation.
- The wind turbine installation project was completed by A2 Sea and Seacore companies. Among them, 24 wind turbines in the deep water area were installed by A2 Sea Company, while Seacore Company installed 6 wind turbines in the shallow water area.
- The installation of wind turbines is completed by Vestas Celtic at their factory in Campel-town, and the wind turbines and blades are pre-assembled by Vestas Celtic at SLP Engineering's Lowestoft dock.

5.2.3 Three-Blade Installation Construction

The three-blade installation involves assembling the three blades and the hub of the wind turbine onshore to form the rotor, without connecting it to the nacelle. During transportation, the angle at which the blades are positioned must be adjusted to effectively utilize deck space, allowing for a reasonable arrangement on the deck. During offshore installation, the nacelle is first hoisted onto the tower, and then the pre-assembled rotor is directly hoisted onto the nacelle. This approach reduces the steps required for positioning and docking the blades during offshore installation, thereby lowering the difficulty of offshore construction, as shown in Fig. 5.28.

Fig. 5.28 Installation of three blades

1. **Installation method of three blade wind turbine**

   (1) The first installation procedure for a three bladed wind turbine includes the tower, nacelle, generator, hub, and pre assembly with three blades. The hub and three blades are assembled separately on land, (referred to as impellers), as shown in Fig. 5.29.

Installation process:

- The tower is lifted separately in sections using one or two cranes;
- Lift the engine room and generator separately;
- Lift the assembled impeller with the main and auxiliary cranes;
- Complete the hoisting of wind turbines.

This installation approach necessitates some components being pre assembled on land, while the impeller takes an extensive assembly period on land, imposing high demands on dock resources. When lifting the impeller, it is required that the lifting capacity of the lifting crane meet the minimum lifting capacity of not less than 300 tons at the hub height. During assembling, the impeller size are substantial, and the lifting vessel is a self- elevating platform. We need to develop impeller assembly fixtures and require wind turbines to have impeller lifting points and corresponding lifting equipment. The onshore assembly of impellers can eliminate the necessity for single blade offshore lifting, resulting in marginally enhanced offshore lifting efficiency and average lifting expenses. At present, this lifting method is mostly used for onshore wind farm operations.

(2) The second installation method for a three-blade wind turbine includes: a tower + nacelle pre-assembled with the generator + pre-assembled hub with three blades. The assembly of the nacelle and generator, as well as the assembly of the hub with the three blades (referred to as the rotor), is completed on land, as shown in Fig. 5.30.

Installation process:

- The tower is lifted separately in sections using one or two cranes;

(a) Installation of nacelle (b) Installation of impellers

Fig. 5.29 Installation method of the first three bladed wind turbine

Fig. 5.30 Installation method of the second type of three bladed wind turbine

- Lifting the assembled engine room and generator;
- Lift the impeller with the main and auxiliary cranes;
- Complete the hoisting of wind turbines.

The installation approach necessitates that the majority of components be pre-assembled onshore, resulting in a high demand for dock resources. The engine room and generator are lifted simultaneously, and the lifting capacity of the lifting crane is required to meet a minimum lifting capacity of not less than 400 tons at the hub height. The assembled impeller has a considerable size and can be lifted by a self elevating platform barge or a self-propelled installation ship. We need to develop assembly support fixtures for the engine room and generator, in addition to impeller assembly fixtures. At the same time, it is required that the wind turbine has assembly lifting points for the engine room and generator, and impeller lifting devices.

Point and corresponding lifting equipment. Due to the large amount of assembly work on land, the efficiency of offshore lifting is relatively high, and the lifting cost is appropriate. This installation method is widely used, such as Goldwind 2 The hoisting of 5 MW wind turbines for the Rudong intertidal zone project and Repower for the Alpha Ventus project.

(3) The third installation method for a three bladed wind turbine: tower pre assembly and pre assembly of the nacelle, generator, hub, and three blades. The tower is assembled on land and the nacelle and generator, as well as the blades and hub, are assembled separately.

Installation process:

- The tower is lifted separately as a whole;
- Lifting the assembled engine room and generator;
- Lift the impeller.

This installation method requires the large components to be pre-assembled onshore, and the assembly of the tower on land requires the establishment of a foundation platform, so imposing significant depends on dock resources. The concurrent lifting of the engine room and generator, the simultaneous lifting of the hub and generator, and the simultaneous lifting of the assembled tower all require the lifting capacity of the crane to meet a minimum threshold of 800 tons. The lifting vessel that can be utilized is a self elevating platform barge. This installation technique requires the design of assembly support fixtures for the engine room and generator, as well as impeller assembly support fixtures. At the same time, the wind turbine is required to have lifting points and corresponding tooling lifting equipment for the assembly and hoisting of the engine room, generator, and impeller. Due to the lack of assembly of tower and impeller, the lifting efficiency is high, yet the expense of onshore assembly cost is substantial.

2. **Typical cases**

The hoisting of wind turbines in Nysted offshore wind farm (see Fig. 5.31) is a typical three bladed installation construction.

Examples of technological applications. The Nysted offshore wind farm comprises 72 wind turbines, organized into 8 parallel arrays, each containing 9 turbines per row. The turbines were supplied by Bonus (a Danish company acquired by Siemens). In 2004, Bonus's maximum-capacity turbine was 2.3 MW, and the total farm capacity reached 165.6 MW. Each tower stands at 69 m—nearly 10% shorter than onshore counterparts.

The design of the substation was completed by ISC Consulting Engineering Company. The design of high-voltage cables and communication systems was completed by the technical department of SEAS company. The lifting work for laying the foundation is completed by a crane ship. Installation of Ship Parameters are summarized in Table 5.5.

5.2.4 Comparison of Advantages and Disadvantages of Various Installation Methods

The comparison of advantages and disadvantages of various installation methods and their application situations are shown in Table 5.6.

Following the completion of the basic installation, the lifting of wind turbines begins. Multiple factors influence the lifting of offshore wind turbines, primarily encompassing the extent of pre assembly of wind turbine components on the dock

Fig. 5.31 Nysted offshore wind farm wind turbine hoisting

Table 5.5 Installation ship parameters

Type	Construction water depth (m)	Lifting capacity (ton)
Self-propelled installation vessel	24	100
Tripod crane	> 100	3300
Tripod crane	> 100	2000

(or land), transportation (ship) and installation (ship) conditions, the quantity and parameters of wind turbines installed (such as quality and external dimensions), distance from the port, and meteorological conditions of the installation sea area. The above factors are not independent and have a certain mutual influence on each other. The degree of pre assembly at the dock or offshore distance affects the selection of transport ships and lifting ships, which in turn affects the determination of lifting methods. The quantity and parameters of wind turbines installed will also affect the transportation and lifting methods. The meteorological conditions in the installation area affect the selection of lifting methods. The lifting method of wind turbines determines the maximum lifting capacity of the elevator, which in turn determines the minimum lifting capacity requirements for transportation and installation ships. Among the above influencing factors, the degree of pre assembly of wind turbine

Table 5.6 Comparison and application of advantages and disadvantages of various installation methods

Installation method	Advantage	Disadvantages	Application situation
Single blade style	(1) Less occupied dock and deck area, flexible layout (2) Horizontal lifting is stable and easy to operate (3) Low lifting cost	(1) High requirements for lifting equipment (2) Multiple lifting operations (3) High requirements for transmission system design (horizontal installation) (4) There are many offshore construction procedures, a large amount of high-altitude operations, and a small operating space	Siemens 3.6 MW, 4 MW, 6 MW offshore wind turbines
Rabbit ear style	(1) Less lifting frequency (2) Low lifting cost (3) The lifting capacity of the crane is average	(1) Require high weather and wind speed conditions (2) The assembled blades occupy the dock and deck area, making them unsuitable for long-distance transportation (3) There are many offshore construction procedures, a large amount of high-altitude operations, and a small operating space	Alstom Offshore wind turbines
Three blade style	(1) Less lifting frequency (2) Reduce offshore operation time	(1) Require high weather and wind speed conditions (2) The assembled blades occupy a large area of the dock and deck	Goldwind 2.5 MW Rudong Project
Integral	(1) Shorten the construction period of the project (2) High work efficiency	(1) Need a large crane vessel with high carrying capacity (2) The design of lifting equipment is complex (3) High installation accuracy (4) High requirements for assembly sites such as docks	Huarui Donghai Bridge Project

components at the dock (or onshore) and the choice of lifting vessel significantly influence the lifting plan.

5.3 Construction of Offshore Wind Turbine Cable

5.3.1 Selection of Submarine Cables and Submarine Cable Accessories

1. **Underwater optoelectronic composite cable**

At present, the submarine cables predominantly used for offshore wind farms are submarine optoelectronic composite cables. It is a fiber optic wind turbine integrated with optical communication function and reinforced structure added to submarine power cables, which serves dual functions: power transmission and fiber optic information transmission, utilized for monitoring and controlling wind turbine systems, and can completely substitute submarine cables and optical cables laid on the same route. In the event of a malfunction, the entire cable needs to be replaced. The capacity and offshore distance of offshore wind farms vary, and the transmission and transformation system methods and voltage level requirements used are also different. The submarine cables used in offshore wind farms can be classified into two categories: internal array submarine cables and output submarine cables.

(1) Internal array submarine cable

The internal array submarine cable primarily connects various wind turbines of offshore wind farms and converge them into the submarine cable of the offshore boosting station. Because the output voltage of wind turbines is relatively low (less than 1 kV, usually 690 V), if the 690 V voltage output by wind turbines is directly transmitted over long distances, it will cause a significant loss of electrical energy. Hence, it is necessary to use the transformer of the wind turbine itself to raise the voltage to medium voltage (such as 35 kV), and then connect it to other wind turbines and offshore substations using internal array submarine cables. In addition to the conductor and insulation layer (usually cross-linked polyethylene or ethylene propylene rubber), the internal array submarine cable also needs to be designed with a radial water blocking layer on the outside to protect the insulation layer from damage caused by water ingress and ensure insulation strength. The water blocking layer can be made of copper (aluminum)/plastic composite tape or lead sheath. Due to the low electrical strength requirements of the insulation layer, medium voltage cables (such as 35 kV submarine cables) generally do not use metal sheaths, but use simpler sheaths. In addition, another crucial structural component of submarine cables is armor. It can provide mechanical protection and tension stability for submarine cables. For each offshore project, the armoring of the submarine cables should be designed based on the tension stability of each area in the cable routing plan (which is influenced by the weight of the cable hangers, different installation methods, and additional dynamic forces generated by the laying vessel), as well as the forms of external hazards and protection requirements (to prevent external threats from installation tools, fishing gear, and anchoring equipment), among other factors.

The maximum length of medium-voltage AC submarine cables is normally less than 30 km.

(2) Output submarine cable

Output submarine cable is the transmission of electrical energy generated by offshore wind farms to submarine cables. Due to differences in wind farm capacity, offshore distance, and wind farm conditions, the transmission and transformation forms of offshore wind farms and the required output submarine cable voltage levels are also different. For offshore wind farms with large capacity and offshore booster stations, high-voltage submarine cables (such as 110/220 V) are usually used for the output submarine cables. The maximum length of high-voltage (greater than 110 kV) submarine cables is generally less than 150 km. For offshore wind farms that are close to shore, have small capacity, and do not require offshore booster stations, medium voltage (usually 35 kV) submarine cables are usually directly used as the output submarine cables of the wind farm.

Submarine cables are generally not standardized items; most projects have different design specifications for cables due to varying site conditions and requirements. In actual projects, numerous investors will endeavor to utilize established standard cables as much as possible to minimize redesign and type testing expenses.

2. **Submarine cable accessories**

Whether it is the internal array submarine cable of the wind farm or the high-voltage output submarine cable of the wind farm, some accessories for submarine cable connection, protection, etc. will also be used in the production, installation, laying, and maintenance process, mainly including submarine cable installation joints, terminals, and other accessories, such as traction heads, anchoring devices, bending protection devices, J-shaped pipe sealing devices, riser submarine cable centering devices, etc.

(1) Submarine cable installation joint

Submarine cable installation joints are mainly categorized into factory joints and submarine installation joints according to their purposes. Factory joints mainly serve to connect semi-finished underwater cables before armor installation, or to connect cables that have been cut off due to production accidents; Subsea installation joints are mainly used for cable installation, connection, and maintenance. The installation joints of submarine cables can be categorized into flexible joints and rigid joints according to the manner of joining.

- Flexible joint. A flexible junction refers to a "field joint" of underwater cables that is formed on the cable laying vessel or in the beach area. Depending on the architecture of the submarine cable and the joint, the fabrication of flexible joints for nearshore cables normally takes 1–10 days. The ends of the two undersea cables need to be positioned aboard the vessel in advance before forming the flexible joint. Due to the lengthy on-site construction process, prefabricated stiff joints are typically employed during the installation and maintenance of offshore wind farms. Flexible joints can be used to link long lengths of cable to successive delivery lengths nearshore.

- Rigid joint. There exists a significant distinction between rigid joints and flexible joints. Rigid joints possess a rigid exterior, with stainless steel pipes as the predominant material utilized. The outer metal protective box is not only the connection point at the end of the submarine cable, but also used for external protection of the internal submarine cable joints. The electrical connections within the external metal protective box may utilize prefabricated connectors as electrical components of the cable. Prefabricated joints generally employ elastic shrink tubing with a semi conductive layer and an insulation layer to bridge the gap between the insulation layers of the cables on both sides. Prefabricated joints offer numerous advantages, including short assembly time, suitability for various conductor connection methods, and the ability to conduct pretests in factories, resulting in their extensive utilization.

The external metallic protective box, owing to its superior mechanical strength, excellent sealing capabilities, and resistance to seawater corrosion, is also suitable for prefabricated joints in submarine cables. The external metal protective box is composed of multiple components, which are connected by welding or bolts. The exterior metal protection box can be mechanically connected to the cable armor or lead sheath using clamping flanges, or welded to completely cover the prefabricated joint, so ensuring safety and waterproofing. To prevent sharp bending and damage to the submarine cable at the transition between the metal protective box outside the rigid joint and the submarine cable armor layer, a bending restriction device made of conical rubber sleeve is required to encase the submarine cable at the cable outlet of the steel protective box.

(2) Traction head

Traction head is a component used for connecting steel wire rope and cable conductor, facilitating the transmission of traction force to the cable conductor, preventing excessive traction force on the cable sheath and insulation layer, and ensuring that the cable remains intact during the laying process. Benefits of the traction head: used in conjunction with traction net cover; Simple structure and easy installation; Equipped with automatic spin reduction function.

(3) Anchor device

The anchoring device is a fixing mechanism installed on the outer platform of the wind turbine, used to secure the vertical suspension cable section and bear the self-weight of the suspension cable section. The anchoring device binds the cable armor layer to withstand mechanical loads through a carefully crafted flange structure. The cable core with lead or plastic protective cover is connected upwards to the cable terminal inside the wind turbine through an anchoring device. The anchoring device is simple to install and uses high-quality anti-corrosion materials to ensure the safety of the submarine cable.

(4) Bending protection device

Like all flexible products, submarine cables having the shortest bending radius, and areas with discontinuous bending strength are prone to excessive bending and fatigue. When threading cables into rigid joint metal shells, fixed device entrances, or cable sealing floating device entrances, a bending limiter should be used to avoid excessive or repeated bending of the cable. The bending limiter can be made of polymer or metal, and is fitted onto the cable to gradually increase the bending stiffness at the connection, ensuring the minimum bending radius of the cable, preventing excessive bending, and reducing fatigue and damage caused by marine environments. However, it should be mentioned, though, that if the bending limiter is not designed properly, the cable may also experience sharp bending at the end of the bending limiter.

(5) J-shaped pipe sealing device

The J-shaped pipe sealing device is used at the J-shaped pipe connection of offshore wind turbine foundations to prevent seawater from entering the interior of the J-shaped pipe and corroding it.

(6) Vertical pipe submarine cable centering device

The vertical pipe submarine cable centering device is used at the connection between the submarine cable protection pipe and the submarine cable to prevent fatigue and damage to the submarine cable caused by friction with the pipe mouth due to the swinging of the submarine cable. It is widely used in European offshore wind power projects and is not currently comely used in the construction of offshore wind farms in China.

5.3.2 Construction Equipment for Laying Submarine Cables

A cable-laying vessel, also known as a cable installation ship, is a specialist ship used for laying undersea cables on the seabed and underwater, and it can also function as a cable repair vessel. The bow of the vessel has a unique design and is equipped with multiple large-diameter cable guiding sheaves. Typically, the vessel needs to be moved during the cable laying process. There are many propulsion options for moving the vessel: (1) employing two tugboats to maneuver the cable-laying vessel from the sides; (2) relying on the working anchor of the cable-laying vessel; and (3) self-propulsion of the cable-laying vessel.

A tugboat, is a vessel used to towships, wooden rafts, or help large vessels in entering or leaving ports, docking or rescuing ships in distress at sea. The tugboat does not have a cargo hold for loading goods, and the hull is not large. It is equipped with a high-power propulsion engine and towing equipment, and has the characteristics of being small and strong. Tugboards are divided into ocean tugboats, port tugboats, and inland tugboats. A tugboat can tow a ship in a row like a locomotive towing a train carriage, or it can tow a barge from both sides of the ship. Tugboats can also tow large ships, and several small tugboats can simultaneously tow a 10,000 tons large ship, allowing the large ship to smoothly enter and exit the port, adjust its position, and enter and exit the dock.

Cable laying ships often require auxiliary equipment such as anti-twist frames, chutes, cable laying machines, meters, tensioners, tension testers, water inlet tanks, water inlet angle indicators, and equipment for underwater cable burial and burial detection during operation.

Twisting frame: Due to the twisting phenomenon of cables during coiling and loading, torsional stress is generated. Therefore, before cable laying construction, the torsional stress of the cables must be released to restore their natural state for laying.

Chute: Chute is a channel for cable laying, which keeps the cable in a certain shape and requirements. It is laid into the water device through various tools, including semi-circular chute, inclined chute, transition chute, and flat chute.

Meter counter: a calculation device used to measure the length of laid cables.

Tensioner: During the process of laying submarine cables, it maintains a certain tension to prevent the cables from slipping off the cable laying vessel due to their own gravity, currents, and wind and waves, which may cause damage to the cables during the laying process.

Tension tester: used to detect the tension that cables are subjected to during construction.

Water entry trough and water entry angle indicator: The water entry trough is the last piece of equipment on the cable-laying vessel before the cable enters the water. As the cable passes through the water entry trough, it is subjected to both the pressure exerted by the cable and the lateral forces generated by the vessel's heading and tugging. Therefore, the design and installation of the water entry trough must be sturdy and reliable.

Cable burying machine: mainly used for pulling, braking cables and burying cables during laying construction. Due to different geological conditions of the project, cable laying machines can be of excavation and burial type, knife plow type, water flushing type, etc. according to specific requirements.

Cable burial detection equipment: mainly used to monitor the posture of cable burial machines on the seabed, excavation depth, GPS position, high-pressure water pump pressure, and cable tension.

Other auxiliary equipment: including traction winches forceable landing, inflatable floating tires, cable tugboats, ropes for tying cables, wooden pry bars, etc. When moving ships, tugboats, anchor boats, and self-propelled ships are also required.

5.3.3 Construction Technology for Laying Submarine Cables

1. **Installation process of submarine cable laying**

The installation strategy for submarine cables in offshore wind farms will be meticulously analyzed and developed based on different hydrogeological conditions, offshore distances, etc., but the overall installation process is basically similar. In addition to preliminary preparations such as land and sea surveys, equipment debugging, sea surface pretreatment, and environmental protection, the submarine cable laying work mainly includes submarine cable factory loading, submarine cable landing section laying, submarine cable placement, submarine cable protection, installation of side submarine cables for wind turbines, marking of submarine cable buoys, and final experiments and production operations.

(1) Preliminary survey and preparation for sea sweeping

The submarine cable route survey is the cornerstone of submarine cable system engineering design and construction. It needs a first field inspection of the coastal topography, landforms, and features, as well as consultations with departments pertaining to the ocean, navigation, geology, hydrology, shipping, fisheries, aquaculture, construction planning, military, and communications. This method helps gather numerous relevant materials linked to the submarine cable project for comparative study, leading to the preliminary identification of the cable landing places and routing plan.

Subsequently, advanced technology and equipment are applied for the submarine cable route surveying to pick safe and dependable landing locations and routes, enabling an economically viable cable laying technical scheme that guarantees the safety and stability of undersea cable communications. Finally, building is carried out based on the surveyed route, utilizing the required optical cables and construction blueprints.

The goal of seabed clearance is to further remove any residual obstructions such as fishing nets. This is normally done with a towing vessel that pulls seabed clearing anchors, making at least one round trip along the path. During seabed clearing, the entry angle of the seabed cleaning anchors should not exceed 30°, and the vessel's speed should be controlled within 6 knots, sailing along the intended course at a low pace. If obstructions are encountered, divers will proceed underwater to clear them. Seabed clearing can be grouped into numerous methods based on the operation type, including anchor towing, ROV (remotely operated vehicle) clearing, and the use of sonar and multi-beam instruments.

(2) Submarine cable loading

If the submarine cable manufacturer is not in proximity to the dock, it is necessary to first use a large flatbed truck to transport the submarine cable reel to the dock, which is expensive. Therefore, many factories will be built near the dock and can directly load cables or ships at the dock. When installing cables, the cable laying ship docks and fixes them, and the submarine cables are transported to the cable laying ship through a cable wharf and coiled on a fixed cable compartment or cable platform. In addition, submarine cables can also be loaded onto pallets or spools. The submarine cable reel can load several hundred meters of armored submarine cables, and there are also oversized submarine cable reels that can load 1–2 km of submarine cables at a time. The cable reel for loading armored submarine cables weighs 30–50 tons and can be directly hoisted onto the deck of the cable laying vessel using a suitable crane. Only when the submarine cable is armored can it be coiled, and the submarine cable will twist 360° in the submarine cable reel. Therefore, the submarine cable reel needs to have the minimum radius. This radius should generally not be less than 60 times the diameter of the submarine cable in order to absorb the torque generated by each turn of the submarine cable. The correct diameter of the submarine cable reel depends on the pitch of the submarine cable armor, the lower layer of the armor wire, and other design parameters. When coiling the cable, attention should be paid to the direction. Only when twisting the cable in the direction where the armor is released can the submarine cable be coiled. If the winding direction is incorrect, it will cause the submarine cable to knot and loop, resulting in cable winding failure.

The array cable length of offshore wind farms is usually 400–800 m, and there are generally two cable supply methods to choose from according to the actual situation on site.

- Use pre-cut cable reels to install submarine cables. Factories need to know the correct length of submarine cables in advance, which is often difficult. Therefore, a large margin of installation length needs to be left, resulting in a large number of abandoned short submarine cables during installation. At the same time, empty submarine cable reels need to be recycled and processed, which is time-consuming and costly. And customers may also request factory testing for each submarine cable, increasing the number and cost of testing.
- Use a large-sized submarine cable loop or turntable for shipping, and cut it according to installation requirements on site. This will result in fewer discarded submarine cables and no need to recycle submarine cable reels. However, there are high requirements for equipment in the factory and on site, and the appropriate cable supply method should be selected based on the supplier and the actual situation on site during project implementation.

(3) Laying of submarine cable landing section

There are usually two methods for laying submarine cables from the sea.

- Lay a fixed anchor at the designated location, connect it with a steel wire rope, and connect both ends to the buried anchor and the conduit towing head. The jacket laying ship approaches the platform, and after the steel cable is tensioned, the jacket is lowered by moving the ship forward.
- Mount a guide pulley at the base of the conduit rack, and connect the steel cable of the winch for the laying vessel to the conduit towing head after passing through the underwater guide pulley. The pipe laying ship is far away from the platform.

When the steel cable of the winch is tensioned, the strain is set and the pipe laying ship is anchored in place. The jacket is lowered by the winch. When the conduit is dragged to the predetermined position, a connecting steel cable is used to replace the winch cable, retract the winch cable, and begin normal pipe laying operations.

(4) Cable laying and protection

As early as the early 1980s and 1990s, cables laid on the seabed were often exposed in the marine environment and easily damaged. By the mid-1990s, underwater burial machines emerged and direct burial protection (referred to as direct burial) was gradually adopted in long-distance cable laying. Nowadays, almost all submarine cables are protected by direct burial or external covering. Direct burial usually includes several forms such as flushing method, plowing method, cutting method, pre-excavation method, etc. Coverage is usually protected by artificial coverings.

① Directly buried.

- Jet Plowing approach: This approach uses a burial plow equipped with nozzles on its excavation segment. There are primarily two types of jet burial plows. One variant is a shoe-style jetting machine mounted to the side of the vessel. The height of the shoe is changeable based on water depth, and both the bottom and root of the shoe are equipped with nozzles. The high-pressure water shot out creates a trench on the seabed, directing the optical cable and interrupter into the trench through the guide rails inside the shoe, which naturally backfills and buries the cable. The other type is a plow-style burying equipment that is towed by a steel cable. The nozzles are situated below the burial arm, directing the optical cable into the burial arm's channel and guiding it to the trench. The maximum burial depth of a jetting burial plow can reach up to 10 m, and the burial speed is proportional to the bottom geology and burial depth, often ranging from 1 to 15 m per minute. The operational water depth is often less than 60 m. The equipment used in the jet plowing method is normally non-powered and principally includes the shoe-type jetting machine, jet plow, and underwater ROV, although there are also powered choices such as underwater robots equipped with their own thrusters.
- Knife plowing method. This method entails a cable laying vessel using a towing cable to pull and inter a plow. A knife plow installed at the tail of the plow excavates a trench on the seabed and then buries the submarine cable and relay. The cable laying ship used in the knife plow method is relatively large, requiring fewer auxiliary ships, strong wind resistance, fast construction speed, and the burial depth is determined by the knife plow. The burial depth can also be adjusted through the control system. The buried plow is equipped with hydraulic devices on the main body to adjust the depth and width of the operation, as well as various sensors and television cameras to monitor the posture, obstacles, and fiber optic cable introduction of the buried plow. There is also an onboard signal processing device for collecting various data from the television camera and various components of the buried plow main body. During burial, the burying plow is controlled and

powered by the monitoring room on the cable laying ship and the power supply cable. At present, the latest buried plow in the world has a burial capacity of over 4 m and a maximum operating water depth of 2000 m.

- Cutting method. Before installation the submarine cable, it is necessary to use cutting wheels, cutting chains, or other professional mechanical crushers to dig trenches along the predetermined route. This is suitable for seabed rock layers or hard soil layers that are not conducive to flushing and plowing methods, and can chop up the hard seabed surface. This burial construction method is slow and costly, and is generally suitable for short distance submarine cable laying.
- Pre-excavation method. First, use a backhoe dredger to excavate the trench, then lay the submarine cable in the trench with a cable laying ship, and fill the trench with a dredger. This method can float the submarine cable above the trench or directly lay it in the trench through an air bag. The dredger is responsible for clearing the silt in waterways and rivers, so that other ships can pass through smoothly.

② Covering protection laws.

When it is impossible to avoid rock sections, when undersea cables intersect with other jackets, when the seabed soil layeris too thin, and when large ships are not allowed to accessing the corresponding sea area, artificial coverage protection is usually used to protect the laid submarine cables. The artificial coverings that can be used include half cast iron pipes, concrete cover plates, cement fillers, and stone throwing. Cast iron pipes can be operated underwater by divers. Concrete cover plates and cement fillers can be lowered from sea level to corresponding positions, and more advanced ones can be operated by underwater robots. Stone throwing protection is a commonly used method, in which professional ships carry crushed stones along submarine jackets and throw them down from the bottom of the ship. A contemporary and sophisticated approach is to use flexible stone dropping pipes to throw the crushed stones down, with corresponding control and monitoring during the laying process.

(5) Installation and fixation of submarine cables

Method for connecting the submarine cable on the side of the wind turbine: Pull the lead out of the J-shaped tube, connect the lead head to one end of the submarine cable to be connected, slowly place the submarine cable into the seabed, and slowly drag the lead at the outlet of the upper end of the J-shaped tube to gradually approach the lower inlet of the J-shaped tube and enter the tube. Continue pulling the lead until the submarine cable is pulled out.

The connection method between the opposite extremity of the wind turbine submarine cable and the wind farm submarine cable: After the installation personnel salvage the two submarine cable ends from the bottom of the water, they connect the two ends on the cable laying ship and slowly sink the connected submarine cable into the seabed through a traction device.

(6) Protection during submarine cable installation

When burying or repairing submarine cables and encountering construction interruptions caused by bad weather, sea conditions, or other factors, it is often necessary to place the submarine cables back on the seabed and set up buoys for positioning and marking. In addition, during the process of laying submarine cables, in order to prevent damage from other construction equipment, buoys are often set at a certain distance around the submarine cables for warning.

(7) Protection of submarine cables after installation

After the installation of submarine cables, continuing maintenance is necessary, particularly to notify individuals to the presence of cables, emphasizing that contact may be dangerous and advising caution to avoid them. Common approaches used for this purpose include:

1. Erecting warning signs in the beach area;
2. Informing relevant departments, such as pipeline operations, maritime bureaus, fisheries management, and meteorological and hydrological agencies, about the location of the submarine cables, and notifying fishermen and sailors;
3. Monitoring the navigation of vessels near the cable route and conducting patrols at sea and in the air.

(8) Submarine cable experiment and production operation

After the submarine cable is put into operation, it is necessary to conduct a DC high-potential voltage test on the submarine cable. This test is the main method to evaluate the insulation performance and overvoltage resistance of the submarine cable, and can effectively detect local defects such as mechanical damage and medium moisture in the oil filled submarine cable. The DC withstand voltage test is a destructive test, and under the action of DC high voltage, the insulation is still prone to damage, which affects the stable operation in the future. Therefore, it is necessary to study the insulation faults that may occur during the DC withstand voltage test. The basic principle of DC withstand voltage test is to apply DC voltage to the main insulation of the submarine cable. This DC voltage requirement is higher than the normal operating voltage of the submarine cable. Maintain the DC voltage for a period of time and keep the voltage value as constant as possible. If the submarine cable sample can withstand such a high DC voltage without breakdown, it can be judged that the submarine cable passes the requirements. When the submarine cable passes the test and meets the requirements, it can be put into operation.

2. **Construction technology of intertidal submarine cables**

The intertidal zone is a zone located between the high and low tide levels of a large tide, which is submerged and exposed with the rise and fall of tides. According to preliminary estimates, the average annual wind speed in the intertidal zone can reach

6–7 m/s, indicating the availability of usable wind energy. In the intertidal zone, the wind turbines of wind farms need to be connected by submarine cables, with a burial depth of at least 2 m. The intertidal zone is characterized by high tide flooding and low tide exposure, with a water depth of only 3–5 m during high tide. These characteristics determine that common offshore cable construction techniques are not suitable for intertidal zones. Below are several common methods currently used.

(1) Cable laying ship construction

During high water levels, a shallow-draft barge is towed to the starting position of the submarine cable route using a cable. The cable guide cage located at the stern of the barge is then lowered into the water. One end of the guide cage is connected to the barge, while the other end is equipped with a hydraulic burial machine. The submarine cable is released from the guide cage onto the beach surface using the guide torsion frame on the barge.

Once the hydraulic burial machine is activated, high-pressure water is sprayed to impact the beach surface, burying the cable to the required depth, such as 2 m. Some cable-laying vessels are also equipped with crawler-type cable laying machines, which can be used to lay various specifications of submarine cables with diameters ranging from 27 to 100 mm. The crawler-type cable laying machine, also known as a linear cable laying machine, uses a hydraulic motor connected to a reduction gear to synchronously drive two chain belts that grip the cable. The submarine cable is then fed into the seabed through a chute arranged at the stern of the vessel.

(2) Excavation method construction

Dredging boats or manual trenching can be used, and the excavation depth is determined by the burial depth. After the excavation is completed, the submarine cable is directly laid down, and the cable is naturally backfilled and buried using power such as currents. The excavation method has low cost and is widely used in intertidal areas. During construction, the line can be laid at high tide in the intertidal zone, and manual excavation can be carried out after low tide. Amphibious excavators can also be used for excavation, and floating equipment can be used to lay cables after filling the trench with water.

Amphibious excavator is a multi-purpose excavator suitable for land, swamp soft ground and shallow water working environments. The walking device adopts a multi-body ship type floating box structure and sealed box shaped track plates, which can safely walk and operate in mud and water surfaces.

(3) Bathtub construction method

Firstly, use a dredging ship to excavate a ditch that can make the cable laying ship float. The width and depth of the ditch should ensure that the cable laying ship can remain floating during the ebb and flow of tides. The cable laying ship has been following behind the dredging ship to lay submarine cables, and at the same time, a floating ship has also been following behind the cable laying ship. During the laying process, the dredging vessel uses a pump to transport the excavated mud and sand slurry through a specialized jacket to the tail floating vessel, which fills the mud and sand into the trench where the cables are laid.

5.4 Summary of This Chapter

This chapter provides a detailed explanation of the foundation structure construction, installation construction, and submarine cable construction of offshore wind turbines, and provides a comprehensive introduction to the possible contents involved in offshore wind turbine construction.

References

1. Bi YX, Zhao SQ, Sun Q, et al. Beijing: China Water Resources and Hydropower Press; 2017.
2. Lin YF. Integrated analysis and design of support structure and foundation for offshore wind turbines. Beijing: Machinery Industry Press; 2020.
3. Yuan Y, Yan HM, Zhang G, et al. Offshore wind power generation technology. Nanjing: Hohai University Press; 2014.

Chapter 6
Offshore Wind Turbine Maintenance and Operation

The levelized cost of energy (LCOE) for offshore wind power is heavily influenced by Operation and Maintenance (O&M) expenditures. Modern offshore wind O&M is based on reliability theory and depends on wind turbine fault state monitoring and diagnosis. With the large-scale and clustered development of offshore wind farms, offshore wind turbine O&M has become more important. Offshore wind turbine dependability metrics, fault categorization, condition monitoring, fault diagnosis, and on-site O&M will be covered in this chapter.

6.1 Dependability Measures

The probability that a component or system will perform its functions under predetermined conditions within a particular time frame is called reliability. Reliability metrics quantify product reliability. Two categories of metrics exist: probability indicators like reliability, unreliability, failure probability density, and failure rate, and life indicators like mean life, reliable life, life variance, median life, and characteristic life. Other reliability indicators for repairable products include mean time to repair, maintainability, and repair rate.

6.1.1 Indicators of Probability

1. **Reliability**

The probability that a product will execute its function under established conditions within a specified time is called reliability. The reliability function, $R(t)$, is a time-dependent function.

M. Ge et al., *Offshore Wind Turbine Technology*,
https://doi.org/10.1007/978-981-96-2175-0_6

$$R(t) = P(T > t) \tag{6.1}$$

The formula uses T for product lifespan and t for stipulated time. $t = 0$ production time results in $R(0)$=1 when t =0 and $R(\infty)$=0 when $t = \infty$.

2. **Unreliability**

The probability that a product would fail to perform its function under predetermined conditions within a specified time is called unreliability. It is sometimes called cumulative failure probability and is a function of time, F(t).

$$F(t) = P(T \leq t) \tag{6.2}$$

The formula uses T for product lifespan and t for stipulated time. $F(0)$=0 when t =0 and $F(\infty)$= 1 when $t = \infty$.

3. **Probability of Failure Density**

The failure probability density refers to the derivative of the cumulative failure probability $F(t)$ with respect to time. The term refers to the likelihood that the duration of the product's existence will occur inside a specific time period around t, specifically the probability of the product malfunctioning during the unit time at time t.

$$f(t) = \frac{dF(t)}{dt} = F'(t) \tag{6.3}$$

4. **Failure Rate**

A product's failure rate describes the likelihood of failure within a unit of time following operation till time t. Essentially, it is the ratio of failures per unit of time after t to the number of unfailed items at t.

$$\lambda(t) = \lim_{\Delta t \to 0} \frac{P(t < T \leq t + \Delta t | T > t)}{\Delta t} \tag{6.4}$$

This equation uses the product's typical operating time, T.
Thus, we can conclude

$$R(t) = e^{-\int_0^t \lambda(t)dt} \tag{6.5}$$

$$F(t) = 1 - e^{-\int_0^t \lambda(t)dt} \tag{6.6}$$

Figure 6.1 shows the "bathtub curve," the link between the failure rate of most items and time, as established by analyzing enormous amounts of product usage and testing data. Product failures have three stages: early, random, and wear-out.

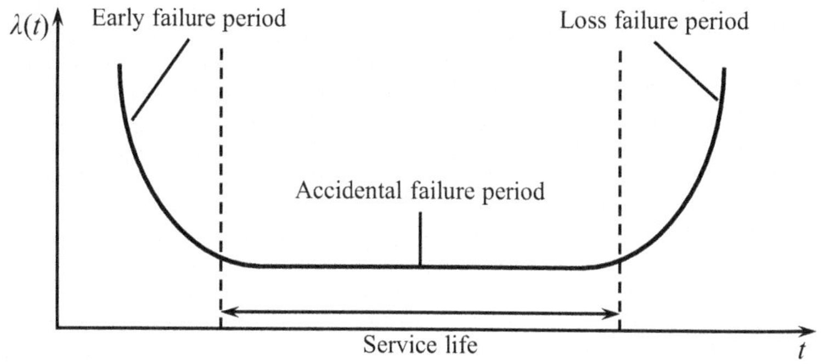

Fig. 6.1 Bathtub curve

(1) The Early Failure Period

Early failure is defined by a high failure rate that diminishes fast as product operation continues. Design, raw materials, and manufacturing faults including defective materials or incorrect assembly and modifications cause these failures. Increasing raw material and process inspections, product quality management, and reliability screenings can shorten this phase and reduce product failure.

(2) Occasionally Fail

This is the random failure or stable operation phase. Low and consistent failure rate, generally approximated as a constant. This is the ideal product use period. Random factors like quality faults, material weaknesses, ambient circumstances, and inappropriate use cause most failures during this time. The product should be carefully maintained and cared for during this phase to extend its lifespan.

(3) Duration of Failure

This stage has a steadily rising failure rate, leading to product failure and scrapping. Wear, tiredness, and age are the main culprits. Situation-specific remedial steps should be done during this time. When aging components fail a system, they can be replaced. Products having lower lifespans can be maintained and replaced preventively. Consider these solutions when designing systems.

The bathtub curve shows product failure probabilities with time and analyzes key distinctions at each step. This gives a theoretical foundation and methodologies for establishing product usage and maintenance strategies that are phased, integrated, and coordinated and ongoing.

6.1.2 Lifetime Indicators

Sometimes life characteristics describe system reliability. Typical indicators are mean and reliable life.

1. **Life expectancy**

Mean life indicates a product's average performance, and life variance shows product lifetime dispersion. The mean life of a batch of items using lifetime data $t_1, t_2, \cdots, t_n$, can be determined from n samples:

$$\theta = \frac{1}{n}\sum_{i=1}^{n} t_i \tag{6.7}$$

The projected life variance is:

$$\sigma^2 = \sqrt{\frac{1}{n}\sum_{i=1}^{n}(t_i - \theta)^2} \tag{6.8}$$

Mean life, also known as Mean Time to Failure (MTTF), is the average amount of time that a non-repairable product lasts from the moment it is started until it breaks down. Mean life for repairable products is the average time between failures (MTBF).

2. **Reliable Life**

The working time t_r corresponding to the reliability R is called reliable life. Here, r refers to the reliability level, satisfying $R(t_r) = r$. When the life distribution follows an exponential distribution, it is expressed as:

$$R(t_r) = r = e^{-\lambda t_r} \tag{6.9}$$

It can be derived as:

$$t_r = -\ln(r)/\lambda \tag{6.10}$$

For exponential distributions, the reliable life under any reliability level can be calculated from the mean life. Product median life is $r = 0.5$ dependability. When the product's life equals the mean, it's called characteristic. Figure 6.2 shows reliable, median, and typical life.

6.1.3 Repairable Product Maintenance Indicators

Maintainability, maintainability degree, and mean time to repair indicate repairable product reliability.

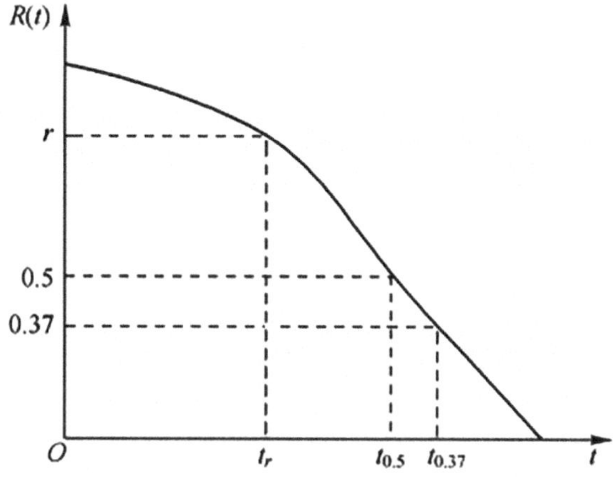

Fig. 6.2 Reliable life, median life and characteristic life

1. **Maintainability and Level**

Repairability is the ability to sustain or restore a function by making repairs under certain conditions, times, procedures, methods, and other constraints. Degree of repair is the probability of maintaining or restoring a condition that fulfils a specified function when repaired under specified conditions. Maintainability $M(t)$ varies with time. Since the repair time τ is arbitrary, the maintainability level is the likelihood of finishing the repair in t. it is expressed as:

$$M(t) = P(\tau \leq t) \tag{6.11}$$

The formula uses t for specified repair time and τ for actual repair time. The likelihood of completing the repair within time t, starting from $t = 0$, is represented by $M(t)$. The cumulative probability function is non-decreasing over time $M(t)$, the probability density function of the repair time t, is also the probability of the product being repaired in a unit of time.

$$m(t) = \frac{dM(t)}{dt} \tag{6.12}$$

2. **Repair Rate**

Repair Rate measures the likelihood of repairing a product that has not been fixed by time t in a unit of time after t. As a function of time, it is denoted as $\mu(t)$. Repair rate and maintainability relate:

$$M(t) = 1 - \exp\left[- \int_{0}^{t} \mu(t) \right] dt \tag{6.13}$$

3. **Mean Time to Repair**

Mean Time to Repair (MTTR) is the average time needed to repair and restore a product from failure to operation. Product characteristics determine this average. MTTR comprises failure identification, isolation, component replacement, calibration, and testing, but not management or logistical support delays. The mean repair time in discrete states is:

$$MTTR = \frac{1}{n} \sum_{i=1}^{n} \Delta t_i \tag{6.14}$$

The formula defines Δt_i as the total repair time for n faults. The mean repair time for continuous distribution is:

$$MTTR = \int_{0}^{\infty} tm(t)dt = \int_{0}^{\infty} G(t)dt \tag{6.15}$$

The irreparability:

$$G(t) = 1 - M(t) = P(\tau > t) \tag{6.16}$$

Use reliability metrics (characteristic quantities) to represent system reliability changes and assess operational characteristics at different reliability stages to base wind turbine component operation and maintenance on this information. This method controls wind turbine operational and maintenance costs and boosts wind farm efficiency.

6.2 Classifying Offshore Wind Turbine Faults

Wind turbines transform wind energy into electricity. Because wind energy fluctuates and is unknown, wind turbine components endure large load variations. Due to the rigorous operational environment, wind turbines often fail. The failure behavior and reasons of each component are listed by location.

6.2.1 Blade and Pitch System Failures

1. **Blade Failures**

Blade icing, cracking, fatigue, surface pitting, and lightning strikes are the principal blade defects [1].

(1) Blade Ice

As demonstrated in Fig. 6.3, blade icing in cold climates severely impacts wind turbine systems. Low temperatures and excessive humidity can cause wind turbine blades to ice. When the turbine is off, the blade can freeze.

Icing affects blade aerodynamic drag and mass distribution, influencing balance and wind turbine power production. Research shows that blade icing can cut wind turbine power output by 25% and halve the power standard deviation. Blade icing increases the blades' bending moment and fatigue load, which accelerates wind turbine fatigue damage, shortens blade lifespan, and may change their natural frequency [1].

(2) Blade Cracks

Wind and sand regularly abrade a wind turbine's resin covering, which can cause uneven bonding and stress concentration sites causing blade breakage after a few years. Blade bonding regions can also break naturally from frequent bending and torsional vibrations. The blade's windward ridge cracks most easily and sustains the most damage. Epoxy glue expands and contracts with seasonal temperature changes, shrinking and splitting the blade's skin. When the skin splits, rainwater can enter the blade, diminishing structural material moisture resistance and causing blade fractures [1].

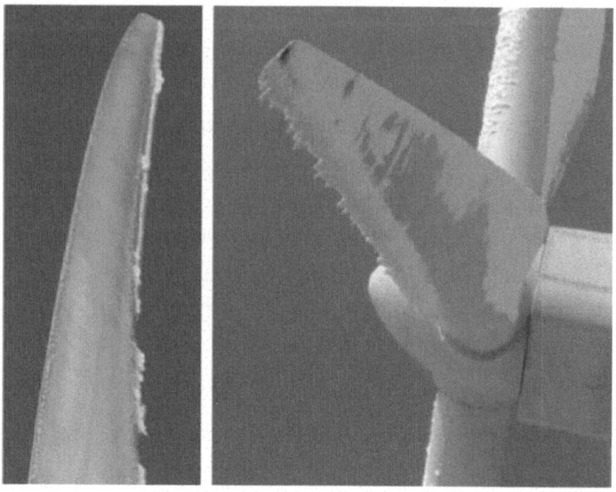

Fig. 6.3 Blade ice

(3) Blade Fatigue

Cyclic fatigue stresses can cause blade fatigue failure in wind turbines. The blades are prone to fatigue damage due to the large number of cycles, even though these fatigue stresses rarely exceed the material's strength limitations. Wind turbine installation angle or pitch angle deviations can produce blade imbalance, limiting blade fatigue stress limit [1].

(4) Lightning Strike and Blade Surface Pitting Faults

Pitting happens when the blade's protective covering erodes. Due to salt spray, high temperatures, and high humidity, the blade's gel layer might deteriorate and produce microscopic pits. Pitting promotes blade deterioration, causing through-holes and loss of protective coating. These through-holes let water into the blade during rainy seasons, increasing interior moisture and reducing lightning protection. Lightning strikes are more likely [1].

2. **Pitching System Issues**

Pitch angle differences, motor failures, gear failures, bearing troubles, and pitch control system failures are the main pitch system flaws [2].

6.2.2 Main Bearing Failures

1. **Fault Types**

The hostile working environment, complicated and changeable operating circumstances, extensive operational cycles, and irregular running times make wind turbine main bearings prone to failure. Human errors like faulty assembly, oil sealing, and maintenance and non-human causes like overloading or prolonged operation can cause these failures. Before diagnosing bearing troubles, you must grasp fault kinds and causes. Fatigue wear, fatigue spalling, corrosion, plastic deformation, galling, fracture, and cage deformation damage are the main rolling bearing faults [3].

2. **Fault Qualities**

External excitation causes rolling bearings to vibrate, whether healthy or injured. The bearing's natural characteristic frequency is the vibration frequency that causes resonance. At manufacture, each rolling bearing component's inherent characteristic frequency is set for its lifetime. Only the bearing's construction affects this frequency [3].

3. **Fault Mechanism**

Failures in rolling bearings have many reasons. Failures can result from fatigue spalling, seal failure, low bearing clearance, and insufficient lubrication. The type of fault can help identify the source, enabling preventive measures or prompt repair. A third of rolling bearing failures are due to typical fatigue spalling, another third to premature failure due to insufficient lubrication, and the other third to contamination or faulty installation. Incorrect installation or contamination can cause poor lubrication and heat-induced failure. Thus, rolling bearing lifespan depends on lubrication. Improper lubrication causes much bearing damage in the following ways [4]:

(1) Over-lubrication

Increased lubricant injection during operation can cause oil overheating owing to stirring. As temperature rises, grease oxidation accelerates dramatically. Leakage might result from over-lubrication.

(2) Under-lubrication

Insufficient lubrication fails to dissipate bearing heat during operation, causing rapid rolling bearing wear.

(3) Bad Lubricant Choice

Performance characteristics including density, viscosity, and others differ in lubricants. Different bearings require different lubricants, and using the wrong one can reduce bearing life.

(4) Lubricant Degradation

A precise blend of additives and basic oils, lubricant's density, viscosity, and additive proportions fluctuate over time, causing deterioration and poor lubrication.

(5) Water Pollution

Water-contaminated lubricant struggles to produce an oil coating on the bearing's metal contact surfaces, failing to dissipate heat. Friction increases with oil film degradation, promoting bearing ageing. Additionally, water in the lubricant corrodes all components it contacts, putting the system at risk.

(6) Particle contamination

Hard lubricant particles accelerate bearing wear, surface roughness, and spalling. Extreme wear outside the rolling line creates micropeaks that produce stress concentration spots, accelerating fatigue failure and reducing bearing lifespan.

(7) Bad Installation

Correct rolling bearing installation is essential. Too much clearance causes axial rotor movement, which accelerates bearing root wear; too little causes poor lubrication, heat, and wear.

6.2.3 *Yaw System Faults*

1. **Brake Wear**

A load restraint on the braking system absorbs slight operating changes and stabilises the yaw system, which supports the entire nacelle. Yaw brake disc and pad wear are included [5].

2. **Yaw Gearbox Issues**

Yaw gearbox problems include wear and broken gears [6].

3. **Yaw Gear Issues**

Breaking the large yaw gear or driving gears can cause yaw gear failure. These issues cause strange loudness.

4. **Yaw Bearing Issues**

Yaw bearings use four-point contact ball bearings, and raceways and steel balls deterioration are the main failure modes [7].

5. **Yaw Drive Unit Issues**

Yaw drive, yaw drive reducer, and yaw motor make up the yaw drive unit. Motor damage and abnormal friction are the main issues [7].

6.2.4 *Gearbox*

Gears, rolling bearings, shafts, housing, fasteners, and oil seals make up the gearbox. Its complicated structure works in difficult environments. Common component failures are caused by manufacture and installation faults, load fluctuations, and insufficient lubrication. Gearbox failures account for 10% of rotational machinery defects and 80% of transmission-related faults. Gear failures account for 60%, rolling bearing failures 19%, shaft failures 10%, housing failures 7%, fastener failures 3%, and oil seal failures 1% in the gearbox. The statistics show that gearbox troubles are mostly gear failures, followed by rolling bearing failures. Both sorts of failures account for 80% of gearbox issues. Thus, gear and rolling bearing difficulties cause most gearbox failures [8]. Rolling bearing failure types and causes are similar to those in Sect. 6.2.2 for main bearing problems and will not be repeated here.

1. **Gear Failures**

(1) Failure Types

Gears may fail during operation due to manufacture and installation defects and dynamic impact stresses, losing their functioning. Common gear failures are:Gear surface wear, galling, scoring, contact fatigue, and tooth fracture [8].

(2) Gear Failure Mechanism

Wind turbine gearboxes are usually planetary gearboxes with three or more gear pairs. The operating speed and torque fluctuate greatly across phases, affecting failure rates. Complex operating conditions, abrupt load fluctuations, and harsh environmental elements promote failures. Technical issues, complex loads, poor climatic conditions, and improper lubrication cause wind turbine gearbox gear damage [4].

2. **Drive Shaft Failures**

Gearbox shafts fail mostly from wear, distortion, and fracture. Bearing-shaft motion causes most gearbox shaft wear. The huge gear's concentric shaft in the gearbox has a sluggish rotation speed and high torque, fluctuating with wind speed, making it most prone to deformation problems. Overload, bearing failures, and emergency stops produce shaft fractures [4].

3. **Lubrication Failures**

Normal gear and bearing operation requires lubrication. Gear and bearing failures often result from lubrication concerns. Lubrication problems include oil and system breakdowns [4].

4. **Other Component Failures**

Besides the principal constructions mentioned above, gearbox subsystems including sealing devices, cooling systems, sensors, air filters, and electric heaters can fail [4].

6.2.5 Failures of Generators and Inverters

1. **Generator Failures**

Failures in generators might be electrical or mechanical [9].

(1) Electrical Failures

Generator electrical failures include stator, rotor, and cooling system defects.

- Stator Faults

Stator flaws focus on stator winding overheating, core defects, insulation damage, and grounding concerns. Vibrations, wear, and corrosion, as well as faulty manufacturing, installation, and maintenance, create these defects. They can cause overheating, short circuits, burning, and explosions.

- Rotor Problems

Rotor defects include winding and body faults. Grounding, turn-to-turn short circuits, and damaged winding wires are rotor winding defects. Rotor body problems include bending, cracking, loose components, fatigue, and torsional vibration damage caused by wear, vibration, overheating, or power grid fluctuations.

- Cooling System Issues

Cooling system leaks and blockages occur in the stator and rotor. These can be caused by cooling pipe problems, incorrect installation, vibrations, or cooling medium contaminants. Failures diminish cooling effectiveness, raise temperatures, overheat structures, and damage insulation.

(2) Mechanical Failures

The generator's mechanical construction, comprising the rotor body, support structure, frame, and foundation connections, can fail. Rotor body issues include imbalance, misalignment, cracking, and loose parts. Support structure defects include rolling bearing failures and oil film bearing instability. Frame cracking, foundation loosening, and structural resonance are frame and foundation flaws.

2. **Inverters Failed**

Internal IGBT open-circuit and short-circuit problems cause most wind turbine inverter failures. High voltage or current usually causes short-circuit failures. In a short circuit, the current spikes, requiring an emergency shutdown for system safety. Overcurrent causes component damage, weak connections, and broken drive lines in IGBT open-circuit problems. Open-circuit failures lower power quality, raise harmonics, and force other IGBTs in the inverter to carry larger currents, heating it and possibly burning it out [10, 11].

6.2.6 Other Failures

1. **Brake System Failures**

Wind turbine brake systems include brake callipers, discs, pads, and pipelines. The high-speed gearbox output shaft holds the braking disc. The gearbox's back houses the braking callipers and hydraulic station.

Brake callipers generate braking torque, and any fault can slow braking and reduce torque, affecting turbine stopping. The brake calliper may not release if the solenoid valve is not functional or hydraulic pressure is insufficient. Other issues including air in the system, low safety valve pressure, excessive brake disc-pad clearance, and high oil viscosity might impede brake callipers. Overloading, speed, grease contamination of the brake disc or pads, misaligned or damaged springs, and loose rotating elements can increase braking distance for less torque. A failure between the brake disc and friction pads can wear or destroy the disc, decreasing braking and stopping ability.

These difficulties are caused by excessive impact or unequal materials. Incorrectly setting the brake calliper or installing the brake pads can cause wear, lowering braking torque and system performance. Poor seal ring placement in the brake system can damage seals, leak internally, and wear mechanically, affecting braking performance [12].

2. **Hydraulic System Failures**

Hydraulic system failures include anomalous operations, flow, pressure, noise, overheating, and pollution [12].

A non-functioning proportional valve can prevent system operation, among other causes. Slow system operations can occur from limited flow, high oil viscosity, blocked proportional valves, or old pumps and valve cylinders. Excessive flow or proportional valve feedback failure can cause quick system actions. Poor system operation might come from worn relief valves, air in the oil, worn pumps or cylinders, and faulty proportional valve feedback.

Incorrect Flow: A non-functioning motor, incorrect rotation direction, pump and connection slippage, pump breakage, or lack of oil supply can cause system flow difficulties. Low relief valve settings, leaking or non-closed pressure relief valves, low motor speed, and old pumps and valve cylinders can reduce flow. High relief valve settings, motor speed, or pump selection can increase system flow.

Incorrect Pressure: No flow, no pressure. Low pressure can result from low relief valve settings, leaking or non-closed valves, low motor speed, and worn or broken components. High pressure might result from faulty or high relief valve settings. Uneven pressure can result from worn relief valves, air in the oil, worn pumps or cylinders, and accumulator problems.

Noise: Pumps, relief valves, and motors cause noise. Pump noise can be caused by cavitation, air in the oil, old pumps, and no proportional valve signal. Low relief valve settings, settings near to another valve's, and damaged valve cores or seats can cause loud relief valve noise. Coupling damage, misalignment, or wear can generate motor noise.

Overheating: High system pressure, filthy oil, low oil levels, inappropriate oil viscosity, cooling system failures, or faulty pumps can cause overheating. System overheating might also result from contaminated oil or misaligned relief valves.

Directional valve failures can occur when contamination pushes it into gaps under pressure, damaging the lubricating oil coating or sticking valve cores. Particles in fast-moving oil flow damage internal surfaces, causing pressure control valve failure. Flow control valves' contamination tolerance depends on their throttle opening form. Contamination can also damage gear pumps.

3. **Submarine Cable Breaks**

Submarine cable failures usually include insulation breakdown and ground discharge [13]. Some failures are caused by internal and external forces. Material ageing, electrical or water tree problems, and product flaws are internal issues. Overvoltage breakdown, natural damage, and human-induced damage like metal water-blocking sheath loss are external variables. Biogenic damage, seismic activity, chemical corrosion, and fatigue ageing are natural causes. Drilling, dredging, undersea pipeline and cable installation, fishing equipment, anchor damage, and other human-caused incidents are examples [13].

6.3 Offshore Wind Turbine Condition Monitoring Technology

Condition monitoring is a multidisciplinary engineering technology based on equipment science and information science. Electronics, computer, sensor, modern control theory, modern signal processing, artificial intelligence, network communication, modern design, and testing theories have shaped information science [14]. Figure 6.4 depicts wind turbine condition monitoring.

Sensors around equipment capture signals. After signal conditioning, transmission, and sampling, the signal processing module removes redundant information to retrieve condition characteristic values. These characteristic data are input into the condition identification module to acquire diagnostic results, which are then processed by the monitoring and diagnostic decision-making module to make a full equipment diagnosis. Computers or specialized equipment process signals, identify conditions, monitor, and diagnose.

The following technologies provide condition monitoring and fault diagnosis:

- Equipment dynamics, mathematics, mechanics, and physics are needed to support fault mechanisms. Equipment failure is caused by internal physical and chemical changes called fault mechanisms. Fault mechanism analysis can define condition characteristic parameters, standard symptom groupings, and the optimum diagnosis procedures. Thus, fault mechanism analysis underpins monitoring and diagnosis.
- New sensor and signal acquisition methods are needed for signal perception. Sensors on or near equipment acquire signals. Wind turbine sensor setup is presented in Fig. 6.5. Sensors are chosen to represent equipment condition changes. Sometimes sensor redundancy is considered. Different sensors can catch

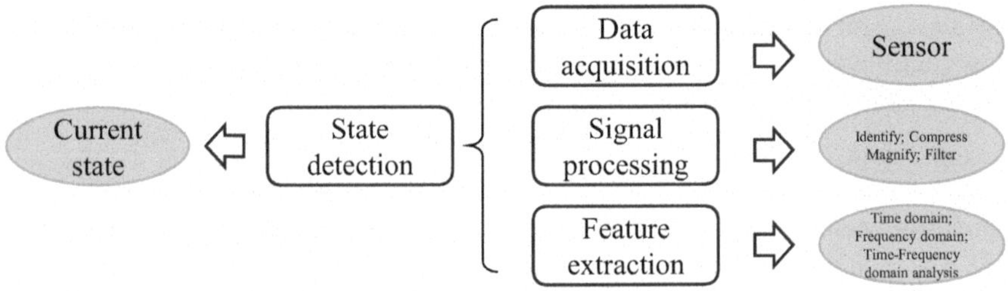

Fig. 6.4 Wind turbine condition monitoring process

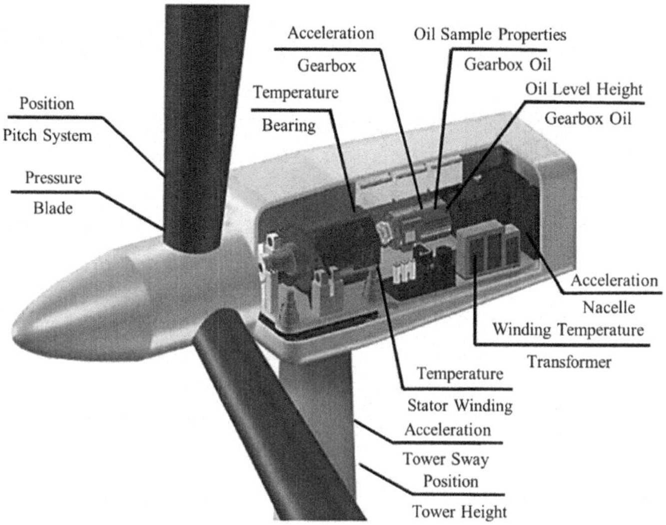

Fig. 6.5 Wind turbine sensor layout diagram

equipment condition changes from different angles to increase diagnostic accuracy. Since sensor data may vary, multi-sensor information fusion technology can combine redundant or complementary data based on certain criteria to provide a unified interpretation or description of the monitored object. This makes the diagnostic system operate better than its parts. Weighted averaging, Kalman filtering, Bayesian estimation, statistical decision theory, evidence reasoning, rough set theory, confidence factor-based production rules, fuzzy logic, neural networks, and fuzzy pattern recognition are basic multi-sensor information fusion methods.

- Classical and modern signal processing are needed for signal transformation. Signal analysis alters signals to help identify and extract important information, estimating and identifying state information. Traditional signal analysis uses filtering and spectrum analysis. Digital filtering, adaptive filtering, and wavelet analysis have considerably improved signal processing content in recent years. FFT, cepstrum, short-time Fourier transform, Wigner distribution, wavelet, local wave, fractal geometry-based, and fuzzy technology-based spectrum analysis methods exist. The last four techniques are new, but their use in equipment diagnostics has advanced defect diagnosis technology.
- Identification and decision-making technology help with condition identification. Diagnostic decision-making, monitoring, and condition identification are combined. Bayesian decision criteria, linear and nonlinear discriminant function pattern recognition methods, time-series model diagnosis based on probabilistic statistics, distance criteria pattern recognition methods, fuzzy diagnosis principles, gray system diagnostic methods, fault tree analysis, wavelet analysis, chaos analysis, and morphological geometry methods are currently being researched and

applied. Artificial intelligence has led to intelligent diagnostic methods including fuzzy logic, expert systems, and neural networks.

6.3.1 Condition Monitoring Signal and Detection Technologies

1. **Vibration**

In wind turbine condition monitoring, vibration is key. Failure severity can be determined by vibration signal amplitude. Gearboxes, generators, main shafts, bearings, and blades usually have vibration sensors. Accelerometers, velocity, and displacement sensors are the basic vibration sensors (Fig. 6.6). Accelerometers can work from 1 Hz to 30 kHz, velocity sensors from 10 Hz to 1 kHz, and displacement sensors from 1 to 100 Hz. Due to their wide frequency range, accelerometers are commonly utilized in condition monitoring. Accelerometer signals include fault-induced gravitational components. Failures from low-frequency wind turbine vibrations are diagnosed via displacement sensors. Using vibration signals, the gearbox, bearings, rotor and blades, generator, tower, and main shaft are monitored. For instance, the vibration signal's 1P frequency component amplitude measures rotor asymmetry.

Installation and use of vibration sensors under ISO 10816 improves fault identification but increases wiring complexity and expense. Vibration sensors on turbine

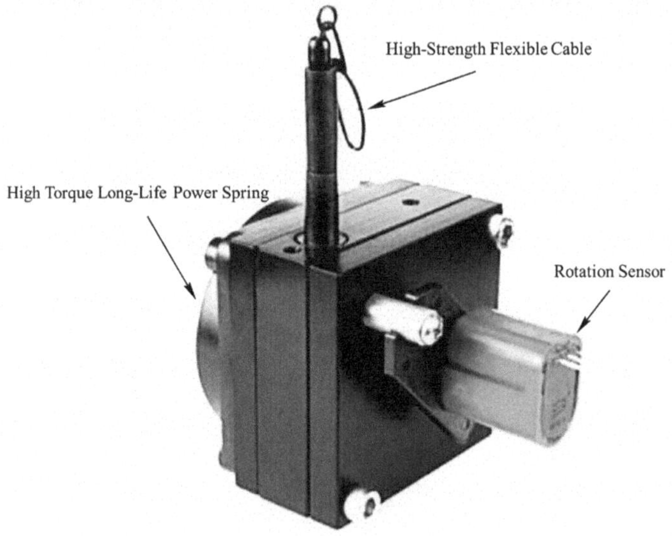

Fig. 6.6 Displacement sensor

surfaces or in turbine components are difficult to install. Sensor and data acquisition device failures can cause control, mechanical, and electrical system failures, increasing maintenance costs and dependability concerns [15, 16].

2. **Sound Emission**

Acoustic emission occurs when a material experiences external loads or internal stresses and releases strain energy as elastic waves. One or more acoustic emission sensors can pinpoint the emission source, and minute structural changes can reveal early structural degradation or faults. Acoustic emission detection, which detects flaws with 50 kHz to 1 MHz vibrations, uses piezoelectric and fiber optic displacement sensors. Acoustic emission waveforms from blade sensors can detect fatigue, stiffness loss, cracks, and increasing surface roughness for early blade failures. Acoustic emission waveform parameters like amplitude and rising time can help anticipate where and when damage will occur, locate it, and track and predict its failure. Gearbox, bearing, shaft, and blade failure detection uses acoustic emission. Acoustic emission signals are excellent for noisy environments due to their high signal-to-noise ratio. Acoustic emission signals run at higher frequencies than vibration signals and can detect early defects or faults before severe harm. This technology requires several sensors, each of which needs a data gathering system for sensing, processing, and transmission, increasing costs and complexity.

3. **Strain**

Strain is essential for structural health monitoring. Common wind turbine condition monitoring sensors are fiber optic strain sensors. Sensors are usually incorporated in blades. Strain signals can indicate blade damage, ice, mass imbalance, and lightning strikes. Tower monitoring uses strain too. Strain measurement has several advantages: (1) it is sensitive to small structural changes, making it useful for early fault detection in wind turbine blades; (2) it can operate at lower sampling rates than vibration, acoustic emission, and electrical signal monitoring; (3) fiber optic strain sensors are passive and do not require external power; and (4) their measurement values do not degrade over time or long transmission distances. Fiber optic strain sensors in wind turbine applications face two challenges: (1) the sensor must always be attached to the monitored material (intrusive); (2) the material's deformation should not exceed the sensor's physical limits, which could cause separation or damage and inaccurate measurements. Strain monitoring, like vibration and acoustic emission monitoring, is expensive, complicated, and requires reliable sensors.

4. **The torque-bending moment relationship**

Rotary torque sensors measure torque, while reaction torque sensors detect bending moment. The rotor, gearbox, and generator usually have torque sensors. Torque can be determined from generator output. Analyzing wind turbine tower torque reveals aerodynamic asymmetry. When a turbine works in skewed wind, blade surface roughness may affect the tower torque spectrum at 1P and 2P frequencies. Rotor flaws can be detected using P amplitude. Torque can detect drivetrain issues. Wind speed changes may generate harmonics that torque sensor readings from the drivetrain can identify while detecting wind turbine generator failures. Rotor imbalances and aerodynamic asymmetry can enhance wind turbine tower second-order bending moments. Torque and bending moment signal-based methods have similar shortcomings to vibration-based methods. Sensors must be installed in more space to measure strain inside components, increasing cost and complexity and reducing reliability. Fault-related torque signals are modulated fault information signals with load-related components. Torque signals for defect diagnostics require more complex signal processing than vibration signals. Thus, torque sensors are rarely employed for wind turbine defect diagnostics compared to vibration sensors.

5. **Temperature**

Temperature measurement is a popular wind turbine condition monitoring method. Under normal operation, wind turbine components should not exceed temperature thresholds. Temperature measurements indicate wind turbine health. Optical high-temperature sensors, resistance temperature detectors (RTDs), and thermocouples are used to monitor gearbox, generator, bearing, and converter temperatures. Monitor gearbox temperature to monitor gearbox and bearing condition, while converters monitor coolant and semiconductor module housing temperature. The monitored area's increased temperatures can be caused by various reasons, making it difficult to determine the cause. If the thermal sensor is positioned on the heat exchanger, a neighboring component failure may raise its temperature. Thus, temperature increases simply suggest a wind turbine malfunction without pinpointing the component. In severe conditions, embedded thermal sensors can disrupt the monitored system and are brittle.

6. **Lube Oil Settings**

Viscosity, oxidation, water, acid, particle count, machine wear, and temperature are the main parameters monitored in lubrication oil. These measurements can be used to monitor oil contamination and degradation, exposing component health and early problem detection. Gearbox, generator, and bearing condition monitoring using lubrication oil monitoring. This condition monitoring method is offline and online. The wind industry uses offline oil condition monitoring, however because samples are obtained frequently (e.g., every six months), it cannot detect defects between sampling intervals. Samples are hard to take while the platform is running. Online oil condition monitoring uses viscometers, liquid level sensors, particle counts, and thermometers to monitor oil condition in real time, eliminating offline issues. However, adding sensors raises monitoring system costs. Not all oil characteristics can be monitored in real time using oil sensors. The many factors that affect oil characteristics under operational settings make real-time measurements from online oil monitoring challenging to understand.

7. **Electrical signals**

Voltage and current data from wind turbine generator and motor terminals can diagnose electrical problems. The amplitude of certain harmonics in electrical signals can detect electrical defects early. Wind turbine components often vibrate due to mechanical or structural problems. Due to electromechanical connection, defective components' vibrations alter generator electrical impulses. Electrical signals carry fault-related information and help identify wind turbine mechanical or structural issues. Existing wind turbine control and protection systems use electrical impulses, removing the need for sensors or data collecting devices. Thus, electrical signal problem identification is inexpensive and simple. Fundamental and harmonic components proportional to nonstationary shaft speed influence fault-related electrical signal components. Nonstationary electrical signal faults require complicated signal processing algorithms to identify.

8. **System SCADA**

Major wind turbine manufacturers install SCADA systems. A typical SCADA system captures statistical properties (e.g., average, maximum, lowest, and standard deviation) of data (e.g., temperature, current, voltage, power, rotor speed, wind speed) every few seconds to 10 min. Sensors in the wind turbine capture data at each interval, and SCADA system outputs can be used for condition monitoring, fault diagnosis, and life forecast using algorithms. SCADA systems are cost-effective for wind turbine condition monitoring and fault diagnostics since they do not require torque or vibration sensors.

Most wind turbine faults' position and mode cannot be determined by frequency or time–frequency studies of SCADA signals. SCADA system signals are mostly used for wind turbine status monitoring and fault prediction using modeling and prediction methodologies.

9. **Non-destructive Testing (NDT) Techniques**

Non-destructive testing (NDT) methods include ultrasonic scanning, infrared thermography, X-ray inspection, and tap testing discover hidden damage in composite materials for wind turbine blade defect diagnostics.

Ultrasonic scanning is the most popular non-destructive testing method used in industry to detect wind turbine blade internal structure problems like delamination. Infrared thermography measures heat distribution on the surface to verify blade health by detecting temperature differences between locations. X-ray inspection penetrates composite materials to show structural changes and wind turbine component faults. Peeling between the skin lamination and the blade's main airfoil can be detected by tap testing, which emits sound waves. The tapping test can determine if the blade's skin laminate and main airfoil are ripping away by emitting sound waves.

NDT methods can detect early defects and track damage propagation, but they require expensive equipment. Most NDT methods are offline, intrusive, and need expert operators and downtime.

10. **Techniques for monitoring corrosion**

Since offshore wind turbine foundations are steel, corrosion monitoring is crucial. Two popular corrosion monitoring methods are:

Resistance Method: A resistance probe thoroughly contacts the medium in the pipeline. Probe parameter changes can cause corrosion-induced weight loss. Corrosion changes cross-sectional area in a fixed-mass, form sensor, changing resistance measurements. Corrosion rate is indicated by this ratio change relative to an uncorroded sensor.

Current Method: This method, also known as the Linear Polarization Resistance (LPR) Method, uses metals' electrochemical polarization in corrosive media. Measure electrochemical response with a three-electrode electrochemical probe in the corrosive environment. Calculating the polarization resistance and using a theoretical conversion coefficient determines the corrosion current (corrosion rate), allowing corrosion speed monitoring.

6.3.2 Wind Power Big Data Platform

Wind power big data platform collects monitoring data from wind farms and integrates functionalities depending on operational management features. It integrates operations, equipment, maintenance, materials, operational status monitoring, safety, and intelligent statistical analysis using real-time data. This system helps companies standardize, refine, digitalize, and integrate wind farm production operations, process management, tracking management, and goal management. It permanently stores integrated data in a data warehouse after filtering and preprocessing. The data warehouse allows the platform to analyze and predict customer operations data marts. Finally, data visualization, mining, and sharing improve operational efficiency and lower costs for wind power management and leadership. Table 6.1 lists wind turbine fault, monitoring, and signal processing signals.

6.4 Technology for Offshore Wind Turbine Fault Diagnosis

Wind turbines have many complicated components with several failure sites. Fault identification is difficult employing additional equipment during maintenance due to the confined internal space. For offshore wind power development, sensor and SCADA-based fault diagnosis technologies enable online fault diagnosis and unmanned wind farm operations, improving wind turbine operation and maintenance.

Table 6.1 Wind turbine component faults, monitoring signals, and signal processing methods

Component	Fault/defect	Monitoring signal	Signal processing method
Rotor	Fatigue Crack Surface roughness Asymmetry Stiffness reduction Deformation	Vibration Acoustic emission Strain Torque Electrical signal Non-destructive testing	Hilbert transform Synchronous sampling FFT Wavelet transform Model-based methods Artificial intelligence
Transmission chain	Gear wear Gear crack Gear fracture	Vibration Acoustic emission Torque Electrical signal Temperature Oil parameters	Hilbert transform Synchronous sampling FFT Statistical methods Model-based methods Artificial intelligence
Bearings	Surface roughness Fatigue, crack, fracture Inner/outer ring Ball Cage	Vibration Acoustic emission Electrical signal Temperature Oil parameters	Hilbert transform Synchronous sampling FFT Model-based methods Artificial intelligence
Main shaft	Corrosion Crack Misalignment, couping fault	Vibration Torque Electrical signal	Synchronous sampling Hilbert transform FFT
Hydraulic system	Oil leakage Valve blockage	Pressure	Threshold comparison
Mechanical brake	Brake disc/pad wear Brake disc crack Hydraulic failure Motor failure	Vibration Temperature Electrical signal	Statistical methods FFY Model-based methods Threshold comparison
Tower	Corrosion Crack Structural damage	Vibration	FFT Model-based methods

(continued)

Table 6.1 (continued)

Component	Fault/defect	Monitoring signal	Signal processing method
Generator	Open circuit/short circuit Insulation damage Imbalance Rotor bar fracture Crankshaft Bearing damage Air gap eccentricity Magnet fault Rotor mass imbalance	Vibration Torque Temperature Oil parameters Electrical signal All related signals	Synchronous sampling Hilbert transform Envelope analysis Statistical methods FFT Wavelet transform Model-based methods Threshold comparison Artificial intelligence
Converter	Capacitor Circuit board Semiconductor	Temperature Electrical signal	Statistical methods Model-based methods Threshold comparison Artificial intelligence
Sensors	Sensor faults Data processing hardware Communication fault Software fault	All related signals	Statistical methods Model-based methods Artificial intelligence
Control system	Sensor faults Actuator faults Controller faults Communication faults Software faults	All related signals	Statistical methods Model-based methods Artificial intelligence

6.4.1 Technical Overview of Offshore Wind Turbine Fault Diagnosis

Forward prediction uses operational patterns to predict defects, while reverse diagnostic uses observed fault behaviors. These methods have produced model-based and signal-based wind turbine failure diagnosis.

Model-based methods create mathematical models of components and induce flaws to monitor responses. This approach relies on model precision and a detailed understanding of turbine operating and fault responses. simulate-based fault diagnosis works well for smaller components but struggles to simulate big systems like wind turbines with many components and strong interaction.

Signal-based methods use sensors to measure component vibrations and diagnose defects. The accuracy of defect feature recognition depends on fault characteristic signal extraction. Transmission chain defects cause wind turbine failures to last longer, according to statistics. Much research has been done in this area.

Transmission chain signals are mostly vibration signals, and classic feature extraction methods fall into three categories:

- Though straightforward and intuitive, time-domain analysis can only detect a few obvious defect types. Noise interference severely inhibits their effectiveness [17].
- Spectrum analysis, inverse spectrum analysis, envelope analysis, etc. are common frequency domain analysis approaches.
- Time and frequency domain wavelet analysis, empirical modal decomposition, and variational modal decomposition are popular.

Feature extraction methods often require a human to identify the defect kind. Technology has advanced to the point that intelligent fault diagnosis can detect the fault type. Intelligent fault diagnosis technique involves feature extraction and fault pattern identification, whose accuracy depends on feature signals and fault pattern recognition. When combined with optimisation algorithms like particle swarm algorithms, gravitational search algorithms, and iteration, intelligent algorithms like support vector machines, neural networks, fuzzy C-mean clustering algorithms, etc. can classify failure modes quickly [18–23].

In real life, working conditions and load changes create non-stationary, variable-frequency signals from rotating drive chain components, which cause most problems. Order analysis is one of the best ways to process non-stationary variable-frequency signals because it isolates rotational speed defect signals. Order analysis and feature extraction can handle non-stationary variable-frequency signals without speed changes [24–26]. Early fault maintenance is cheap and prevents problem escalation, lowering operational and maintenance expenses. Early fault features rarely appear, and feature signals are weak and noisy. These signals can be denoised using maximum correlated kurtosis deconvolution [20, 27].

Model-based wind turbine fault diagnostics requires a thorough grasp of component models and fault reactions. Signal-based wind turbine failure diagnosis uses fault processes to identify problem features. Thus, studying fault production mechanisms is crucial for fault diagnosis. A gearbox is used to introduce wind turbine problem detection technology from a failure mechanism standpoint.

6.4.2 Torsional Vibration Model and Natural Characteristics Analysis of Wind Turbine Gear Transmission System

1. **Wind Turbine Gear Transmission System Torsional Vibration Equation**

The lumped parameter method simplifies the wind turbine's continuous gear transmission system into a discrete system to create a vibration model, or analogous dynamic model, as shown in Fig. 6.7.

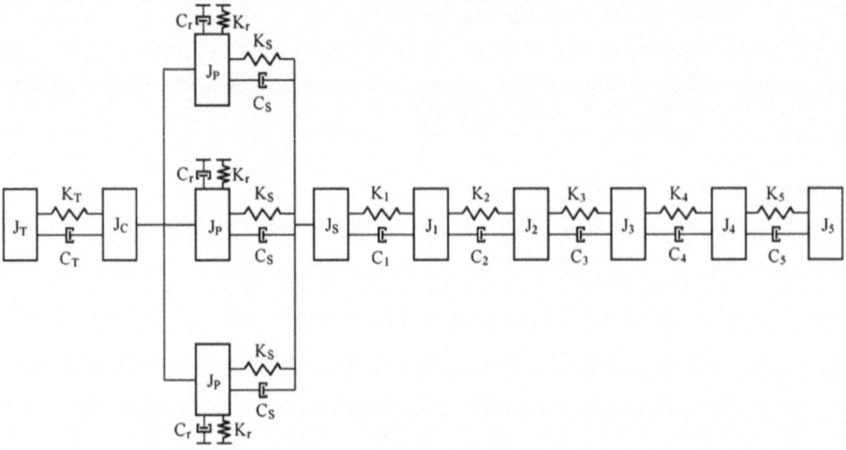

Fig. 6.7 Equivalent dynamic model of wind turbine gear transmission system

Vibration in multistage gear trains generally includes radial, axial and torsional vibration. Studies demonstrate that torsional vibration is the system's main vibration and excites radial and axial vibrations. In this chapter, only the system's torsional vibration is considered and a pure model is created. Using Newton's second law, the wind turbine gear transmission system's motion differential equation may be simplified:

$$J\ddot{\theta} + C\dot{\theta} + K\theta = P \tag{6.17}$$

The equation's angular displacement, velocity, and acceleration matrices are θ, $\dot{\theta}$ and $\ddot{\theta}$. J is the moment of inertia matrix, K the torsional stiffness matrix, C the damping matrix, and P the internal and external excitation matrix.

2. **Torsional Vibration Model-Based Natural Characteristic Analysis**

An important dynamic property of a vibration system is its natural frequency and mode shape. The system's structure, size, shape, and other elements determine these characteristics, which do not alter when vibrating [28]. The equation of motion for undamped free vibration with no external force and negligible damping is:

$$J\ddot{\theta} + K\theta = 0 \tag{6.18}$$

Assuming the solution of the equation is:

$$\theta_i = A_i \sin(\omega_i + \varphi)i = (1, 2, \ldots, n) \tag{6.19}$$

Equation (6.18) can then be rewritten as:

$$\left(K - \omega^2 J\right)A = 0 \tag{6.20}$$

The condition for a non-zero solution is that the determinant of the coefficient matrix is zero, so:

$$\left|K - \omega^2 J\right| = 0 \tag{6.21}$$

The characteristic Eq. (6.21) represents the system's natural frequency (eigenfrequency) and non-zero solution, with ω indicating the eigenvalue. A is the eigenvector for mode shape ω. The structure's spatial form when vibrating at its natural frequency is its physical meaning. National standards can determine stiffness, while *Pro/E* can calculate moment of inertia.

6.4.3 Wind Turbine Gear Transmission System Vibration Response Analysis

1. **Wind Turbine Gear Transmission System Excitation**

Gear gearbox systems are elastic mechanical systems that respond dynamically to stimuli. The gear gearbox system's dynamic excitation is internal and external. External excitation mostly involves driving and load torque fluctuations. Gear gearbox systems generate stiffness and error excitation during meshing, unlike conventional vibration systems. Thus, gear gearbox system dynamics research must examine internal excitation-induced dynamic response [28].

(1) Excitation Stiffness

One of the main types of gear meshing excitation is stiffness excitation, where dynamic meshing forces excite the system due to the time-varying comprehensive meshing stiffness.

Most gears overlap during meshing, causing alternating single- and double-tooth meshing. The gear carries the same load in single- and double-tooth meshing zones. In the double-tooth meshing zone, two pairs of gears share the load, reducing elastic deformation and increasing meshing stiffness. Comprehensive meshing stiffness is lower in the single-tooth zone. Since the gear rotates continually, its meshing stiffness alternates between single-tooth and double-tooth meshing.

Spur gears have single- and double-tooth meshing zones throughout. When switching between these zones, the meshing stiffness varies abruptly, causing each set of gears to bear a different load. Dynamic gear excitation results from load variation.

Helical gears also have alternating single-tooth and double-tooth meshing, but since the meshing starts at one end and gradually goes across the tooth surface before exiting at the other, their comprehensive meshing stiffness, though time-varying, does not change step-like.

(2) Error Excitation

The inherent faults in gear production and assembly cause error excitation. Motion, smoothness, and contact accuracy indicators are used to check these errors. A pair of gears with a gearbox ratio of 1 has angular errors when the driving gear rotates uniformly for one revolution.

Long- and short-period angular errors exist. Gear geometric eccentricity causes long-period faults, which do not affect gear smoothness at typical speeds. Short-period errors are mostly caused by pitch and involute tooth profile abnormalities. For fixed-axis gear gearbox, these faults match meshing frequency. Transmission smoothness is mostly affected by short-period faults. Teeth profile variations cause displacement excitation during meshing, whereas pitch discrepancies cause meshing shocks.

2. **Wind Turbine Gear Transmission System Vibration**

The analysis of the vibration response of the wind turbine gear system belongs to the problem of solving the system of ordinary differential equations, which is solved by the Runge–Kutta methods, and the following conclusions are obtained [28]:

(1) Wind Turbine Gear Transmission System Vibration Response Without Error Excitation

The medium-speed gears' time-varying meshing stiffness excitation greatly influences system vibration. In the power spectrum, vibration energy is high at the meshing frequency and its harmonics of each component's medium-speed gears. At the meshing frequency and harmonics of the high-speed stage, the tiny gear vibrates significantly. Small gears have a higher vibration amplitude than large gears in meshing pairs.

(2) Wind Turbine Gear Transmission System Vibration Response with Error Excitation

A Fourier series with the meshing frequency as the fundamental wave represents gear error excitation. The power spectrum increases because error excitation vibration varies greatly and the limit error is each gear fault's amplitude. Error excitation increases the high-speed shaft gear's vibration amplitude and enriches each component's low-frequency harmonics.

6.4.4 Wind Turbine Gear Transmission System Fault Model

1. **Crack-Fault Gear Meshing Stiffness**

Loaded gear teeth act as cantilever beams during gear gearbox. Mesh gear teeth experience pulsing cyclic bending forces. When stress is too great, bending fatigue can cause cracks that propagate and break teeth. Gear teeth lose bending stiffness when cracked, according to fracture mechanics theory. The teeth' bending stiffness decreases with crack size, reducing gear meshing stiffness.

The magnitude of deflection deformation following loading represents gear meshing stiffness [28]. Gear pairs deflect during loading due to contact deformation at the meshing point, gear tooth bending and shear deformation, and gear root body deformation. To calculate meshing stiffness of gears, use the formula:

$$K = \frac{F_n}{\delta_H + \delta_{T1} + \delta_{T2} + \delta_{A1} + \delta_{A2}} \tag{6.22}$$

where δ_H represents contact deformation, δ_T represents tooth bending and shear deformation, and δ_A represents body deformation at the gear root. Gear cracks mostly effect tooth bending and shear deformation.

Usually, gear cracks occur from tooth root to pitch circle. Figure 6.8 compares meshing stiffness variation curves for gears with fractures in a wind turbine's low-speed shaft gear pair.

The larger gear in the gear pair's meshing stiffness varies with crack length on the left side of the picture. The smaller gear meshing stiffness varies with fracture length on the right. As seen in the figure, the meshing stiffness of the gear pair reduces nonlinearly with crack size in both situations, although the smaller gear crack does so more significantly. When pitch circle cracks emerge, both the smaller and larger gears lose meshing stiffness, but the smaller gear loses more stiffness. Cracks in the smaller gear affect meshing stiffness more in meshing gear pairs.

2. Gear Impact Waveform with Pitting Faults

Pitting or spalling on a gear tooth surface causes a fault impact signal with positive meshing-in and negative meshing-out impacts when the tooth engages and disengages. Label the fault impact signal $\delta(t)$ [28]. Due to gear meshing's periodicity, the fault impact signal matches the faulty gear's rotation frequency. The differential equation of motion for the wind turbine gear transmission system with pitting problems is:

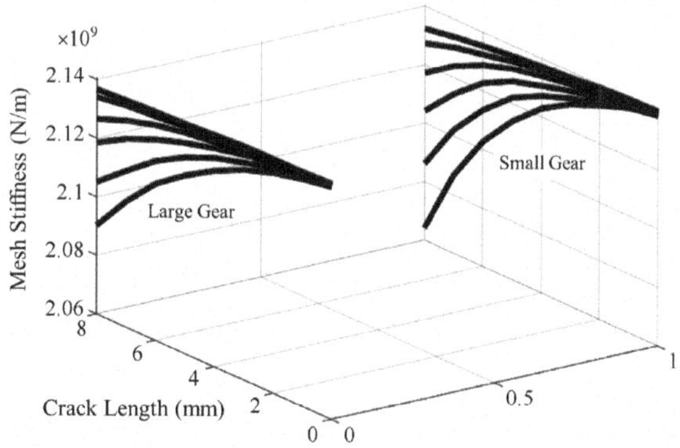

Fig. 6.8 Comparison of meshing stiffness curves of different gears with cracks

$$J\ddot{\theta} + C\dot{\theta} + K\theta = P + \delta(t) \tag{6.23}$$

6.4.5 *Live Wind Turbine Drivetrain Health Assessment Model*

The previous section covered wind turbine gearbox system issues. Fault diagnostic methods have been extensively studied by many scholars. However, most current research focusses on imminent or existing defects, which falls short of wind turbine lifetime management. Real-time health assessment of the wind turbine drivetrain, early fault pre-diagnosis, and component degradation trends allow wind farms to organise maintenance equipment and personnel, procure spare parts, coordinate component replacement, reduce costs and crane preparation time, and ensure safe wind turbine operation while significantly improving operability. Thus, wind turbine drivetrain health checks in real time are crucial. We use the Gaussian Mixture Copula Model to create a real-time wind turbine drivetrain health assessment model. The GMCM outperforms the GMM in data distribution fitting and image segmentation. To monitor health in real time, GMCM model parameters are computed and a degradation index is developed [23].

The GMCM parameter estimation steps are:

- Using a two-dimensional vector as an example, calculate the marginal probability density functions for each dimension separately;
- Based on the two established marginal probability density functions, transform the original data into marginal cumulative distribution functions;
- Using the values of the marginal cumulative distribution functions, calculate the GMCM parameters based on the EM algorithm or the Active-Set algorithm.

Based on the GMCM, and taking a rolling bearing as an example, the input vibration signal of the rolling bearing generates a degradation index (DI) to evaluate the real-time health status of the bearing. The degradation index DI is calculated using the following formula:

$$DI_t = (1 - \alpha)DI_{t-1} + NLLP_t \tag{6.24}$$

where $NLLP_t$ is the GMCM's negative log-likelihood probability at time t and α is the exponentially weighted moving average statistic's smoothing coefficient. α is set at 0.2 to balance the impact of past and current data on the degradation index.

Figure 6.9 depicts the Gaussian Mixture Copula Model-based real-time health evaluation procedure.

And the steps are:

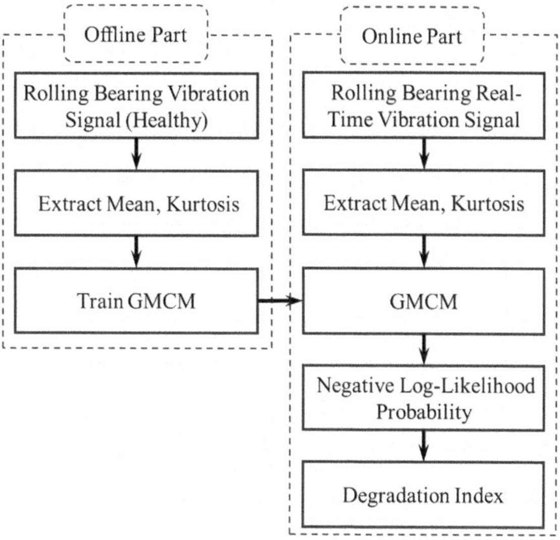

Fig. 6.9 Gaussian Mixture Copula Model-based real-time health evaluation

- Establishing the GMCM: Extract the mean and kurtosis from the rolling bearing's history vibration signals in a healthy state to construct a two-dimensional feature value matrix, calculate the GMCM parameters, and establish the GMCM.
- Calculating the NLLP: Extract the mean and kurtosis from the rolling bearing's real-time vibration data to build a two-dimensional feature value matrix, enter the GMCM, and calculate the NLLP.
- Monitoring the Degradation Index: Calculate the rolling bearing's real-time degradation index DI_t, track its development, and analyse its health using the Formula (6.23).

6.5 Offshore Wind Farm Maintenance and Operations

In extreme conditions, climate, tides, and other factors affect offshore wind farm maintenance. Weather can hinder maintenance vessels from docking at a wind turbine, and even small issues can restrict its availability. The tough marine environment increases preventative maintenance repair costs, raising offshore wind plant operational costs. Operating and maintenance (O&M) costs make up 20%–35% of offshore wind farms' levelized cost of electricity (LCOE), unlike onshore wind farms. Vessels for offshore O&M can account for 73% of maintenance costs. Variable component types require different vessels, staff, and maintenance time. Crane vessels are needed for gearboxes and generators. Replace versus repair solutions for the same component need different amounts of resources [29].

The following analysis will examine offshore wind turbine O&M elements and strategies.

6.5.1 Weather Impacts

Weather greatly affects offshore wind farm operation and maintenance. Storm surges can make vessel navigation unsafe and prevent maintenance staff from doing O&M at any time. Weather is an inherent factor in offshore wind turbine O&M and a major decision variable.

6.5.2 Maintenance Staff

Offshore wind farm maintenance staff must understand offshore wind turbine operation and structure. They must employ computer monitoring systems and understand turbine operational statuses, faults, and problem types. They should also be able to diagnose common faults and their remedies, complete operation and work tickets, and calculate wind turbine capacity factors, utilisation hours, and failure rates statistically.

6.5.3 Spares

Sea repair windows may be brief due to weather, logistics, and other operational constraints. Critical spare parts have extended manufacturing cycles and are connected to asset management, while consumable spare parts are regularly needed and can be maintained as consignment stock. Categories for these parts:

- Blades, gearboxes, generators, hydraulic power units, converters, pitch and yaw motors are crucial.
- Consumable parts: Lights, buttons and control relays, pump motors, filters, lubricants, etc.

6.5.4 Methods of Maintenance

The timing of maintenance divides it into two categories:

(1) Corrective Maintenance (CM)

Corrective maintenance, also known as reactive maintenance, is a passive strategy that remedies a fault after it arises if weather and vessel availability allow. The operations and maintenance (O&M) team must keep replacement parts or rely on equipment manufacturers for fast delivery. Corrective maintenance raises spare part costs and downtime and requires O&M to respond immediately to all wind turbine failures. This method works best for nearshore wind farms where maintenance expenses outweigh repair prices or low-cost, less important equipment. However, far offshore wind farms are less feasible and effective [30].

(2) Preventive Maintenance (PM)

- Time-Based Maintenance (TBM)

It involves using failure patterns to schedule maintenance on equipment at regular intervals, independent of its current condition. 6-month, 1-year, 2-year, and 5-year maintenance intervals are typical. These periodic maintenance plans guide spare component acquisition, storage, and supply. Cleaning, lubrication, adjustments, inspections, overhauls, and replacements are typical maintenance tasks. Table 6.2 lists the primary wind turbine components that need routine maintenance inspections.

Service life and equipment reliability determine maintenance intervals. Periodic replacements are arranged for equipment with high failure rates after a given operational time. Time-based maintenance is only used when failure implications are high and preventative maintenance is feasible and effective. Real-time operating circumstances should inform the periodic maintenance schedule [31].

- Condition-Based Maintenance (CBM)

Condition-based maintenance compares equipment operating circumstances to standard operating conditions and takes relevant maintenance actions. This strategy works for equipment where failure rate is not highly connected with operational time. Although there may not be a clear correlation, most equipment shows indicators of impending failure, and CBM uses technical ways to identify and prevent failure. CBM relies on real-time equipment health monitoring, analysis, early alerts, fault diagnosis, and maintenance scheduling. CBM reduces intervention, improves efficiency, targets maintenance, and lowers costs. It optimises reliability and economic efficiency with big data and IT.

(3) Opportunity Maintenance

- Both preventive and remedial maintenance have pros and cons. Corrective maintenance is unpredictable, while preventive maintenance may over- or under-maintain. When one component fails, opportunistic maintenance allows preventive maintenance on other components [32]. This technique enables proactive maintenance decisions by determining if other components fulfil criteria. Opportunistic maintenance spreads high fixed maintenance expenses by performing numerous tasks at once. It optimises wind farm maintenance by connecting turbines. This method has been tested in other engineering domains, but its use in wind farms and offshore turbines is new.

Table 6.2 Inspection content of main components in wind turbine systems

Component	Inspection content
Rotor blades	Surface damage, cracks, structural discontinuities; inspection on elevator or step equipment, visual inspection or appropriate methods (e.g., tapping, ultrasonic testing); bolt pre-tensioning; damage to lightning protection systems
Transmission system	Oil leakage, abnormal noise, corrosion protection conditions, oil seepage, bolt pre-tensioning; transmission system (relevant oil samples); damage to lightning protection systems
Nacelle and stress and torque transmission components	Corrosion, cracks, abnormal noise, oil seepage, bolt pre-tensioning, damage to lightning protection systems
Air conditioning, dehumidification, and air filtration systems	Functionality, contamination, and dirt accumulation
Hydraulic and pneumatic systems	Damage, leakage, corrosion, functionality
Support structures (tower, substructure, and foundation)	Monopile or lattice structure: corrosion, corrosion protection (e.g., cathodic protection); damage and deformation, cracks, wear, spalling, bolt pre-tensioning, marine growth; seal structure (visual inspection)
Safety equipment, external lighting, sensors, and braking systems	Functionality checks, limit value conditions, damage, wear
Control systems and electrical systems, including substation and grounding systems, environmental monitoring systems	Terminal connections, fastenings, functionality checks, corrosion, dirt accumulation
Helicopter platforms, ship docking areas, protective barriers	Fastenings, functionality, corrosion, cracks, dirt accumulation, damage, and deformation
Emergency shelters	External lighting and emergency shelter equipment, maritime rescue equipment, and backup power (if needed)
Documentation	Completeness of maintenance documentation completed during regular maintenance intervals, test documents, certification documents, environmental monitoring, explanations, etc. Modifications may be made when authorized if necessary

6.5.5 *Options for Offshore Transportation*

1. **Offshore Maintenance Vessels**

Operating and maintaining offshore wind turbines requires specialised ships. These watercraft should operate well in waves, be comfortable to navigate, and dock precisely at low speeds to the turbine base without impact. They must stay in touch with the foundation and convey maintenance staff and equipment to the wind turbine platform safely. Tools, spare parts, and other items should be stored in containers or specialised equipment on the deck. The ship should also make loading and unloading these commodities straightforward and offer overnight berthing for crew.

Based on China's offshore wind power development, maintenance vessels fall into six categories [33]:

- General maintenance vessels;
- Specialized catamaran maintenance vessels;
- High-speed specialized catamarans;
- Maintenance mother ships;
- Accommodation vessels;
- Self-elevating maintenance vessels.

2. **Helicopters**

Helicopters at 300 km/h can reach a wind farm 30 km offshore in 6 min, while maintenance vessels take over an hour. Fast turbine fault resolution can reduce power loss during high wind and electricity generation. Offshore rescue activities require helicopters for safety and rescue [33].

Speed and efficiency distinguish helicopters from maintenance vessels. Helicopters can quickly resolve faults at 2–6-m wave heights. Given cost–benefit, helicopters for offshore maintenance are a good choice.

References

1. Li PF. Research on fault diagnosis and condition evaluation method of wind turbine blade. Beijing: North China Electric Power University; 2015.
2. Li XW. Wind turbine state prediction and pitch system anomaly identification based on data mining. Chongqing: Chongqing University; 2012.
3. Wang TC. Applied research on rolling bearing fault diagnosis of wind turbine based on vibration monitoring. Qinhuangdao: Yanshan University; 2016.
4. Sun XW. Wind turbine gearbox failure modes and impact analysis. Beijing: North China Electric Power University; 2014.
5. Xie L. Acoustic diagnosis of large wind turbine yaw system faults. Beijing: Beijing University of Posts and Telecommunications; 2020.
6. Deng ZH, Li LP, Liu R, et al. Research on wind turbine yaw gearbox fault diagnosis method based on SCADA data feature extraction. J Chin Soc Power Eng. 2021;41(01):43–50.
7. Deng ZH, Li LP, Liu R, et al. Overview of troubleshooting techniques for motor-driven active yaw systems for large wind turbines. Acta Energiae Solaris Sin. 2020;(04):34–41.
8. Dong CH. Wind turbine gearbox operational condition monitoring and fault diagnosis. Beijing: Beijing Jiaotong University; 2017.
9. Xu YJ. Wind turbine generator fault analysis and diagnosis. Beijing: North China Electric Power University; 2013.
10. Guo DJ. Research on wind turbine condition monitoring and fault intelligent diagnosis system. Taiyuan: Shanxi University; 2012.

11. Wang M, Tan YH, He YG, et al. Open-circuit fault diagnosis of permanent magnet direct-drive wind power system converter. Control Eng China. 2018;25(01):50–6.
12. Yang MM. Statistical analysis of failure modes and fault diagnosis of large wind turbines. Beijing: North China Electric Power University; 2009.
13. Qian KM, Lu Y, Zheng M, et al. Methodology for reliability assessment of high voltage XLPE insulated submarine cables for offshore wind farms. Wire Cable 2016;(01):1–6+9.
14. Tchakoua P, Wamkeue R, Ouhrouche M, et al. Wind turbine condition monitoring: state-of-the-art review, new trends, and future challenges. Energies. 2014;7(4):2595–630.
15. Hossain ML, Abu-Siada A, Muyeen SM. Methods for advanced wind turbine condition monitoring and early diagnosis: a literature review. Energies. 2018;11(5):1309.
16. Qiao W, Lu D. A survey on wind turbine condition monitoring and fault diagnosis—part ii: signals and signal processing methods. IEEE Trans Industr Electron. 2015;62(10):6546–57.
17. Zhang B. Direct-drive wind turbine bearing vibration monitoring and fault diagnosis. Taiyuan: Taiyuan University of Technology; 2014.
18. Xin WD. Research on wind turbine drive chain vibration analysis and fault feature extraction methodology. Beijing: North China Electric Power University; 2013.
19. Li Z. Research on intelligent fault diagnosis method of wind turbine drive chain based on unsupervised learning. Beijing: North China Electric Power University; 2016.
20. Zhao HS, Li L. Early fault diagnosis of wind turbine bearings based on MCKD-EMD. Electric Power Autom Equipment. 2017;37(02):29–36.
21. Liu CL, Wu YJ, Zhen CG. Rolling bearing fault diagnosis based on variational modal decomposition and fuzzy C-mean clustering. Proc CSEE. 2015;35(13):3358–65.
22. Su ZQ. Research on early fault diagnosis method of wind turbine drivetrain based on generalized flow learning. Chongqing: Chongqing University; 2015.
23. Xu Q. Research on the state diagnosis method of wind turbine drive chain. Beijing: North China Electric Power University; 2015.
24. He GL. Research on vibration response modulation mechanism and sparse separation method of composite gear drive system. Guangzhou: South China University of Technology; 2015.
25. Wu YJ. Fault diagnosis study of wind turbine drive train based on variational modal decomposition. Beijing: North China Electric Power University; 2016.
26. Li L. Research on wind turbine bearing fault diagnosis based on vibration signals. Beijing: North China Electric Power University; 2017.
27. Lu ZL. Research on early fault diagnosis method of rotating machinery based on variational modal decomposition and optimized multi-kernel support vector machine. Chongqing: Chongqing University; 2016.
28. Long Q. Research on dynamic characteristics and fault diagnosis method of wind turbine gear transmission system. Beijing: North China Electric Power University; 2012.
29. Tavner P. Offshore wind turbines: reliability, availability and maintenance. IET Digital Library; 2012.
30. Fu Y, Xu WX, Liu LJ. Research on operation and maintenance strategies for offshore wind power. J Shanghai Univ Electric Power. 2015;03:219–22.
31. Guo HD. Offshore wind turbine fleet operational condition evaluation and maintenance decision making. Beijing: Beijing Jiaotong University; 2018.
32. Shan GK. Offshore wind power technology. Beijing: China Science Publishing & Media Ltd.; 2020.
33. Zhou H. Offshore wind power operation and maintenance of wind power operation and maintenance ship. Wind Energy. 2017;98(09):48–51.

Chapter 7
Impacts of Offshore Wind Power on the Oceans

The surrounding marine ecology is impacted by offshore wind installations over their whole existence. These consequences include chemical and biological effects on the environment and marine life in addition to physical ones like noise and electromagnetic waves. As a result, this chapter explores the whole range of environmental impacts—physical, chemical, and biological—that offshore wind energy has to offer. Additionally, it presents innovative integrated models for the growth of offshore wind energy, including wave energy, marine ranching, hydrogen production and storage, and other power generation methods.

7.1 Environmental Impacts

Throughout its existence, offshore wind power projects produce a variety of noise and electromagnetic radiation that change the surrounding sea's physical environment in comparison to other marine locations. Offshore wind power, in contrast to onshore wind power, also requires extra care to prevent corrosion. The chemical composition of saltwater in offshore wind farm sites differs from that of other coastal regions since the various degrees to which different corrosion protection techniques leak metals and organic compounds into the water. The research results on the effects of offshore wind power on the physical and chemical environments, both domestic and foreign, are reviewed in the part that follows.

7.1.1 Impacts on the Physical Environment

1. **Noise**

The construction of an offshore wind farm can be divided into preparation, construction, operation and decommissioning phases according to the construction cycle. Each phase generates different types and characteristics of noise [1].

M. Ge et al., *Offshore Wind Turbine Technology*,
https://doi.org/10.1007/978-981-96-2175-0_7

2. **Electromagnetic Radiation Field**

Impacts of electromagnetic radiation from offshore wind power projects are mostly caused by substations, underwater cables, and offshore wind turbines. The location of wind turbines and substations is near the sea surface, where electromagnetic radiation quickly attenuates over medium and has little effect on marine life. Thus, the transmission cables of offshore wind farms are the primary potential source of electromagnetic radiation damaging marine life. Outside the cables, the radiation is mostly magnetic since they are grounded by an insulating layer that shields them from the electric field. As a result, the ferromagnetic components in the shield and the current passing through the cables are the main sources of the magnetic field produced by undersea cables [6].

7.1.2 Impacts on the Chemical Environment

Chemical emissions from offshore wind farms are less harmful to the marine environment as other offshore operations, but they do become more significant as the number of wind farms rises. While there isn't enough evidence to conclude that corrosion protection systems have a negative impact on the marine environment, more research is required to fully understand the potential (long-term) effects of corrosion protection systems, which include offshore wind turbines, offshore booster stations, and high voltage direct current (HVDC) platforms. Corrosion protection systems are a direct source of chemical emissions from all types of offshore structures in wind farms [2].

1. **Metals**

The necessary protection current for the steel structure determines the necessary anode material. The precise computation is dependent on a number of technical specifications and parameters, including the anode distribution, seawater or sediment conditions, size, coating status, and foundation design. Since the anode material differs from offshore wind farm to offshore wind farm and is assumed to be consumed and released into the marine environment over the course of a 25-year life cycle, the chemical emissions from a galvanic anode–cathode protection system will be comparable to the anode material. Consequently, the DNVGL-RP-B401 (2017) computation (one of the recognized standards) can be used to approximate the possible emissions from galvanic anode cathodic protection systems. Table 7.1 Calculations for the required anode materials (zinc, aluminum, and indium) for offshore wind turbine bases with various coatings have been made simpler. For various uncoated casting designs, the aluminum pole mass varies from 13,000 kg (monopile) to 32,000 kg (tripod). The total mass of the aluminum anode will decrease due to the extra coating. In this instance, the needed aluminum anode mass ranges from 6000 kg for a monopod to 10,700 kg for a tripod.

Table 7.1 Necessary anode materials for offshore wind turbine foundations with different coatings (aluminum indium zinc) simplified calculation results

Offshore wind turbine foundation design dimensions	Spar	Derricks	Ductwork	Offshore boosting station conduit racks
Surface area in water (m^2)	850	2500	1800	12,000
Surface area in sediment (m^2)	850	300	180	1200
Total anode material	Spar	Derricks	Ductwork	Offshore boosting station conduit racks
Basic uncoated (kg)	13,000	32,000	22,500	150,000
Base coated[a] (kg)	8000	16,000	11,000	80,000
Base coated[b] (kg)	6000	10,700	7500	50,000

Notes [a] It is assumed that the coating will be completely gone after 15 years, since the standard for coatings assumes only a 15-year service life
[b] Assuming a total coating damage rate of 32% after 25 years (damage factor calculated using DNV GL standard values)

It is important to note that the actual demand must be taken into account for each component of the foundation design and may alter the overall number of anodes. These estimates do not account for any extra elements, such as anode distribution throughout the structure. Using the values from Table 7.1, we can calculate the annual emissions of an offshore wind farm with 80 offshore wind turbines with monopile foundations and an offshore booster station. This wind farm would release 45 tons of aluminum and 2 tons of zinc (assuming a zinc content of 5%). Adding zinc anodes would increase the total annual emissions by a factor of approximately 2.5 (118 tons). Zinc makes up 99% of the emissions from zinc anodes. The amount of anode material required will be greatly decreased by the application of coatings, and emissions from offshore wind farms using aluminum anodes can be lowered to 19–25 tons of aluminum annually. Although the majority of offshore wind turbines do not yet have coatings in the submerged region, emissions from offshore wind farms with aluminum-containing anodes can be decreased to 19–25 tons of aluminum annually. The values are averaged during a 25-year lifespan; the entire base is polarized during the beginning phase, when they are typically higher. Furthermore, offshore booster stations—which typically have larger foundations than offshore wind turbines and require more anode material due to their estimated 0.5 to 1.0 tons of emissions annually—are a common feature of wind farms. The HVDC converter station, which makes up the majority of the offshore wind farm's equipment, needs a lot of anode material (approximately estimated emissions of 5–15 tons/year) because the conduit rack's steel frame alone weighs several thousand tons. As a result, the wind farm's overall metal emissions will rise.

The application of electroplated anodes inside monopiles or tripods, which are used in some offshore wind turbines, is currently being debated due to a number of difficulties. These designs' foundations separate the internal chemical environment of the building from the surrounding water, allowing for little to no chemical exchange. The environment within could alter, leading to a drop in pH, an increase in hydrogen production inside the pile, and a decrease in oxygen concentration (from aerobic to anaerobic circumstances). Research has indicated that these modifications could impact the corrosion process and operation of the galvanic anodes in the pile; for example, reducing the pH could hasten the anode's depletion. In addition, if ventilation is not supplied, there is a risk of explosion from the production of hydrogen gas. The entrance of significant volumes of water with low pH and high metal content should be avoided as it may have a detrimental effect on the marine environment. In addition, the deposits and water inside the pile are rich in aluminum, zinc, and other trace metals released by the galvanic anodes. This ought to be considered, at the very least, when offshore wind turbines are being decommissioned.

In conclusion, during the course of the offshore wind farm, the usage of galvanic anodes on the outside and inside of offshore infrastructures will cause considerable localized metal inputs to the marine environment. Below is a discussion of the anticipated metal releases from galvanic anodes, specifically aluminum, zinc, and indium. It should be noted that other heavy metals, such lead or cadmium, which are similarly released during ingestion, are also continuously present in electroplating anodes.

(1) Aluminum

In the crust of the planet, aluminum ranks as the third most plentiful element. Seawater typically has a low concentration of dissolved aluminum, ranging from 0.008 to 0.68 μg/L in open ocean areas, 0.5 to 0.68 μg/L in coastal waters, and 0.6 to 0.9 μg/L in the North Sea. Owing to seawater's inherent alkalinity (pH = 8.1) and acidity, hydroxide is the most common form of aluminum. In saltwater, aluminum can also react with fluorine and organic materials to generate chelates, or combine with them to make combinations. In more acidic environments, freshwater algae and fish are recognized to be harmful to the ecosystem. For aluminum concentrations in saltwater, there are no established national, international, or environmental quality assessment values. Based on toxicity testing of 11 marine species from different trophic levels, the Soil and Water Resources Institute (SWRI) of CSIRO, Australia, has proposed a No Effect Concentration (NOEC) of 24 μg/L for total aluminum in the Australian seawater guidelines and has shown that, depending on the species investigated, dissolved and/or particulate aluminum causes toxicity. The diatom was the most sensitive marine creature to aluminum concentrations in this study, followed by the oyster, which had NOECs up to 100 μg/L, or roughly four times the NOEC for total aluminum. The observed levels of dissolved aluminum in the marine environment were far lower than these no-effect thresholds.

More precisely, few studies—mostly conducted in port environments or lab settings—have examined the impacts of aluminum emissions from primary cell anodes. The University of Caen study showed that there was no rise in the concentration of aluminum in the water as a result of dilution, but there was a considerable increase in the concentration of aluminum in the sediment close to the anode. High amounts of suspended and settled particulate matter were seen in the tanks used for the tests by other researchers, which used saltwater without any sediment. They speculate that filter-feeding organisms may be connected to the aluminum enrichment in suspended particulate matter. Higher concentrations of dissolved organic carbon and matter would lessen aluminum toxicity for freshwater fish. Additionally, they surmised that the diluting effects of dissolved aluminum in saltwater render it irrelevant in environmental contexts. Through research on aluminum buildup in the purple mussel, Mytilus edulis, in a port setting, the University of Caen has shown that the digestive gland serves as a short- and medium-term storage location for aluminum. Sea urchin bioassays were used to investigate the effects of zinc and aluminum dissolved by galvanic anodes and their related sulphates. The results indicated that the effects of the zinc and aluminum generated by galvanic anodes were negligible. The Australian Commonwealth Scientific and Industrial Organization (CSIO)'s Soil and Water Resources Institute looked at how various diatoms were affected by dissolved and precipitated forms of aluminum. Additionally, they surmised that the diluting effects of dissolved aluminum in saltwater render it irrelevant in environmental contexts. Through research on aluminum buildup in the purple mussel, Mytilus edulis, in a port setting, the University of Caen has shown that the digestive gland serves as a short- and medium-term storage location for aluminum. Sea urchin bioassays were used to investigate the effects of zinc and aluminum dissolved by galvanic anodes and their related sulphates. The results indicated that the effects of the zinc and aluminum generated by galvanic anodes were negligible. The Australian Commonwealth Scientific and Industrial Organization (CSIO)'s Soil and Water Resources Institute looked at how various diatoms were affected by dissolved and precipitated forms of aluminum. They showed that, depending on the type of diatom, either dissolved or precipitated metal could be harmful.

As was previously indicated, the anodes of galvanic connections release a significant amount of aluminum. Since aluminum is greatly diluted in the open ocean and can be released as amorphous aluminum hydroxide or insoluble material, it is unknown if these releases raise the concentration of dissolved aluminum in saltwater. Tank experiments have shown that the content of aluminum in suspended particulate matter can be increased by electroplating anodes. These aluminum-bounded insoluble material and suspended particle matter could be buried in sediments close to offshore wind turbines. It is unclear, therefore, if these flows affect the concentrations of silt and benthic creatures. As no pertinent data are published, it is unknown what the results of detailed discharge studies—which include examinations of metal species and metal complex development on anode surfaces—will be and how they will affect the situation locally or even regionally. It may be challenging to discern between the natural aluminum concentration and the impact of galvanic anodes on the aluminum concentration in sediments because the overall aluminum content of marine sediments is already high due to their genesis in clay minerals. To evaluate the possible ecotoxicological impacts of aluminum hydroxides generated from galvanic anodes, this distinction is crucial. An examination of the morphology of aluminum in sediments is required to comprehend the effects of aluminum emissions from primary cell anodes.

(2) Zinc

In addition to being a common element on Earth, zinc has toxicological effects that vary in dose. Among the marine micronutrients that are widely distributed in the marine environment is zinc. The kind and distribution of zinc dissolved in sediments, suspended sediments, and seawater are significantly influenced by the physical and chemical characteristics of the surrounding conditions. Organic content, redox conditions, pH, and salinity all affect its distribution in the marine environment. According to reports, the concentrations of zinc in the North Sea's open ocean and coastal waters are 0.3 ng/L and 1 μg/L, respectively. The University of Caen conducted the research wherein the quantities of zinc in various North Sea regions, measured offshore, varied between 0.3 and 300 μg/L. The OSPAR environmental monitoring program includes zinc for direct discharges of yearly fluxes and riverine inputs. Riverine inflows to the North Sea were 2689 tons in 2013, making them a significant source. In contrast, offshore wind farms probably emit few emissions. The OSPAR Ecotoxicological Assessment Criteria (EAC) range from 50–500 mg/kg dry weight in sediment and 0.5–5 μg/L in water. These values are merely intended for first evaluation in order to spot possible issues; they have no legal significance. For drinking water, the WHO suggests a limit of 5 mg/L based only on taste parameters.

Because zinc anodes are frequently employed in ships, ballast tanks, and port piles, researchers have looked at the destiny and impacts of zinc anodes in sediments and seawater in a variety of methods. After examining mussels from an offshore gas platform in the Adriatic Sea, the Italian National Research Council talked about galvanic anodes as a possible cause of metal (nickel, zinc, and cadmium) buildup in the mussels, which resulted in a generally mild bioturbation signal. In aluminum-based anode materials, zinc ranks second in terms of abundance, making up 2.5% to 5.75% of the total mass of the anode. Zinc anode tank tests revealed a rise in Zn^{2+} ions in the water, which could precipitate as hydroxides, complexes, or attachments to the suspension. In order to show that the dissolved zinc content only increased at the start of the experiment and then decreased as a result of the adsorption of suspended particulate matter and the dilution effect during the water renewal process, the researchers analyzed the aluminum, zinc, and iron released from the aluminum-based anode material in the seawater tanks. On the anode surface, they noticed a highly hydrated, amorphous white coating. The white layer had a high zinc concentration in the liquid phase and a high aluminum concentration in the solid phase. The toxicity of zinc discharged from the anode on sea urchin sperm and embryos was investigated at the University of Caen. Zinc was less or not hazardous than the aluminum and zinc salts. The effects of exposure to zinc anode discharges on the immune system activity of oysters were also studied. It was found that the immune system activity of long oysters was significantly affected by high concentrations of zinc toxicity, but low concentrations of zinc had no discernible effect on the toxicity of long oysters.

Research indicates that even though the amount of zinc released from offshore wind farms is much less than that of aluminum (roughly 2 tons/year, which was previously used in offshore wind farms with aluminum-zinc-indium anodes), studies are still required to fully comprehend the outcomes and possible effects of this new source of zinc emissions. This is true even though the toxicity of zinc from galvanic anodes is relatively low.

(3) Indium

Just 0.01 to 0.04% of the anode's substance is indium. Electroplated anodes could be a significant new source of indium in marine environments, especially considering the low environmental incidence of only 0.05 ppm in the Earth's crust. Electrical displays, photovoltaic cells, and light-emitting diodes frequently use indium as an indium tin oxide material. Over the past 20 years, industrial production of indium has expanded and is a major source of indium in the environment. Due to issues with experimental analysis and the low indium content, no comprehensive studies were carried out in the majority of the aforementioned emission tests and tank experiments. Sediments from a single flume experiment demonstrating the presence of indium in anodized exposed sediments have only been examined by a small number of researchers. Indium in the marine environment is primarily studied through correlation analysis of geochemical processes, as there are little direct data available. The majority of research on the toxicological consequences of indium has been conducted on industrial workers, and the findings indicate that inhalation of indium can have toxicological effects. The toxic effects of indium (III) on freshwater aquatic creatures were first studied through wastewater discharge due to the growing demand and application of indium in industry. The median fatal concentration of indium (III) was found to be between 6.9 and 21.5 mg/L (LC_{50}). There are indium concentrations of less than 0.6 to 1.6 pmol/kg in the Mediterranean and Atlantic Oceans, and even lower in the Sea of Japan, but as far as we are aware, there is currently no data on indium in the marine environment in Europe.

2. **Organic Matter**

When organic coatings (polyurethanes and epoxies) come into contact with saltwater, they may leak, weather, or lose material, releasing organic materials. Chemicals include adhesives, pigments, fillers, organic modifiers, solvents, and various additives are found in epoxy and polyurethane coatings. Furthermore, the resin is a reaction product of several constituents, and the reaction may result in the production of unidentified compounds. Little research has been done on marine coatings, however leach or release of organic coatings by other processes over their lifetime has been studied for various applications and additives, such as biocides in facade coatings, coatings, and paints. Epoxy resin uses connected to food or drink, or the application of epoxy coatings to aquatic environments for onshore civil engineering infrastructure, are the main areas of research. The ecotoxicity of the organic covering was examined in the aforementioned leaching studies, and it was shown that several exogenous estrogens—like bisphenol A—were present. The specific coating product has a major influence on the outcome and quantity of organic matter emitted. The relationship between the coating hardening time and the organic matter was examined in the first study of its kind, which examined the leach of various organics from a one-component polyurethane corrosion protection system. The findings were in line with previous analyses of this type. To minimize emissions, the coating must be applied and hardened onshore to prevent emissions from incompletely cured coating material being released into the water. It is unclear, though, how frequently the coating is maintained at sea, which could result in extra emissions (particularly during the coating process), though it is challenging to measure these emissions.

(1) Bisphenol A

Epoxy resins frequently begin with bisphenol A (or F) diglycidyl ether, and the coating itself may contain traces of lettable bisphenol A or F after coating. Moreover, bisphenol is occasionally utilized in offshore industrial coatings and can be utilized as a component compound in epoxy resins as free bisphenol A. The evaluation study merely makes the assumption that epoxies without any free BPA components may have a residual BPA level of 10 ppm, which might theoretically leak out and into the environment. Weathering, damage, or the emission of any other residual or replacement bisphenol, like bisphenol F, are not included in this assessment. Furthermore, a number of factors influence the amount of BPA and other chemicals that leak out of coatings. These include variations in BPA concentration in the resin across products, amount used, method of application, brine conditions, exposure to sunlight, and stress corrosion cracking of the coating. Additional research is required as the current data are insufficient to estimate total emissions.

BPA has no natural source, but because it can be used to make various epoxy resins and polymerized plastics, it is already widely present in the environment and is generated in vast quantities. Epoxy resin manufacture accounted for 17–30% of BPA usage in 2008, with an estimated 51 tons of epoxy resins used annually in marine coatings throughout Europe. One important byproduct of epoxy resins is the interaction between epichlorohydrin and bisphenol A (or, alternatively, bisphenol F). Epoxy resins are frequently used to cover offshore wind turbines and ships.

The ability of bisphenol A and bisphenol F to leak from epoxy coatings in various applications, such as drinking water or laboratories, has only been demonstrated in a small number of experiments. Because BPA is a direct source of absorption for humans, the majority of leaching research concentrate on BPA in consumer items, like water bottles. BPA is frequently assumed to have leached into the marine environment from river inputs, microplastics, and marine trash.

Bisphenol A is currently under discussion as an environmental concern. Additionally, various reaction products of epoxy resins are being evaluated, including: 4,4′-isopropylidenediphenol (Bisphenol A) and oligomeric reaction products with 1-chloro-2,3-epoxypropane; formaldehyde and oligomeric reaction products with 1-chloro-2,3-epoxypropane and phenol; 2,3-epoxy-propyl neodecanoate; and 2,3-epoxypropyl o-tolyl ether. Due to the possibility that they could be endocrine disruptors. The Convention for the Protection of the Marine Environment of the North-East Atlantic lists BPA among the compounds of possible concern. The BPA in seawater has a provisional predicted no-impact concentration (PNEC) of 0.15 μg/L, which serves as an environmental quality standard. The initial study also showed how alternatives like BPF have ecotoxicological implications. It has been found that BPA concentrations in the open North and Baltic Seas are generally low, falling below

PNEC limits. BPA was found in river water samples only, not in the open North Sea, according to the German Federal Maritime and Hydrology Agency and the German Elbe Cleaning Task Force (50 ng/L detection limit). The average BPA content in surface water and bottom water of Gdansk Bay, in the Baltic Sea, was 59.2 ng/L and 79.2 ng/L, respectively. Offshore wind turbines could represent a new point source for organic pollutants like bisphenol A in the North Sea and Baltic Sea, as current background concentrations are generally minimal. However, the extent of paint emissions is unknown.

(2) Other organic matter

Screening ingredient information from safety data sheets of various paint products shows that many phenolic compounds, Such as p-tert-butylphenol, octylphenol, nonylphenol or dodecyl phenol (and its isomers) and several other organic compounds (such as xylene and its isomers, ethylbenzene, methyl isobutylketone, trimethylbenzene, ethyl toluene, propyl benzene, butylacetate, 3-ethoxy-propionate, 1-methoxy-2-propanol and its acetate, Benzyl alcohol), often used as a solvent, viscosity regulator or hardening catalyst in coatings, especially the epoxy curing agents of 2,4, 6-triepoxy oligomers are mainly diamines, polyamines, polyamides and imidazole derivatives. The following diamines are used as the most common curing components: m-phenyldimethylamine, isophorone diamine, 1, 5-diamino-2-methylpentane, 1, 2-diamino-cyclohexane, and triethyltetramine. Alcohols and phenolic chemicals are typically avoided in polyurethanes due to hardening issues. The European Chemicals Agency and the Convention for the Protection of the Marine Environment of the North-East Atlantic have already listed nonylphenol and nonylphenol polyoxyethylene ethodes as compounds of environmental concern. These compounds are already limited to uses that come into direct contact with consumers, like textiles. Nonylphenol (0.3μg/L EQS) and 4-nonylphenol (4-nonylphenol, 0.33μg/L PNEC) are two examples of the quality requirements for aquatic environments. 2009 saw an investigation by the University of Montpellier into the prevalence of octylphenol and nonylphenol in the marine environment.

There is relatively little information available regarding the release of chemicals into the environment following coating curing due to the paucity of research on leachable parts in environmental settings. Present research employing more advanced analytical techniques, including non-targeted screening in conjunction with leaching tests or resin analysis, may uncover newly discovered compounds released into the environment by the coating. Research conducted by the German Federal Institute of Hydrology has shown that organic compounds can be leached into water from a single component polyurethane. The length of time that the chemicals are leached and hardened determines how much chemical is released. The substances found in this study are classified into five chemical groups: toluene diisocyanate, p-toluenesulfonamides, carbamic acid (toluenesulfonyl) esters, 4,4′-methylene diphenyl diisocyanate, and p-toluenesulfonic acid derivatives. Deterioration and damage can raise leaching-related emissions from the underlying coating in addition to leaching-related emissions from the entire coating. They can also be a source of particle emissions. As previously stated, the coating should last 15–20 years, and during the first 15 years, the coating surface should fail at a rate of at least 10–20%. This loss of coating materials is attested to by the performance evaluation of anti-corrosion systems for offshore wind transmission platforms by Germany's Meerhan AG, which also notes that the concentration of particles, such as antifouling paint, observed near ship repair facilities, abandoned buildings, and grounded vessels, accounts for the majority of this damage.

7.2 The Biological Impact of Offshore Wind Power

The large-scale development and construction of offshore wind farms will have a certain impact on the Marine ecological environment during the construction and operation periods. A systematic, comprehensive, scientific and objective understanding of the research progress on the biological impact of offshore wind power at home and abroad is of profound significance for promoting the collaborative development of offshore wind power industry and Marine environmental protection [3].

1. **Impacts on Birds**

The noise of pile foundation installation during the construction period, turbine blade collision and electromagnetic field during the operation period of offshore wind farms will affect birds, which are mainly reflected in the following four aspects [3]:

- Affect the behavior of birds, such as the avoidance of birds;
- Affect bird habitat and foraging;
- Wind turbines interfere or form barriers that affect bird migration;
- Death due to collision.

2. **Impacts on Fish**

The impact of offshore wind farms on Marine fish is mainly reflected in the impact of noise and electromagnetic field, which mainly includes the following six aspects [3]:

- Underwater piling noise affects fish behavior and even causes death;
- The construction period affects the growth and development of eggs and young fish;
- Noise generated during the operation phase may cause communication obstruction or disorientation of fish;
- The construction leads to the suspension of sediment and sediment or the leakage of oily waste water pollution of the sea water quality, affecting the life of fish;
- Electromagnetic fields affect the distribution and migration patterns of surrounding fish;
- Electromagnetic fields may affect the early embryonic development of fish.

3. **Impacts on Marine Mammals**

The impact of offshore wind farms on Marine mammals is mainly due to noise, which is mainly reflected in [3]:

- Causes hearing loss in Marine mammals;
- Causing avoidance behavior in Marine mammals;
- It may have an effect on the reproduction of Marine mammals.

Electromagnetic fields generated by wind farms may also have an impact on Marine mammal behavior.

4. **Impacts on Benthos**

The best seabed geological conditions for the construction of wind farms are soft sediment areas. This is also a suitable habitat for many benthic organisms. The effects of offshore wind farms on benthos include [3]:

- During the construction period, it will cause water pollution, change or destroy the habitat;
- During operation period, the sediment composition will be changed and the benthic community structure is affected.

5. **Impacts on Plankton**

Offshore wind farms can impact the Marine hydrological environment, which in turn affects the accumulation of plankton. Artificial seabed building can lead to changes in the seabed environment, resulting in changes in benthic creatures such as invertebrates, which further impact the composition of algae. Plankton in the project waters is subject to short-term, reversible impacts during the construction phase and very little throughout the operation phase. Plankton can have a permanent home at the foundation of a wind farm, which will increase their connectivity [3].

6. **Impacts on Marine Biodiversity**

Some foreign academics have recently expressed the belief that the development of wind farms benefits marine life in some way. When wind farms are completed, they will add to the habitat of marine life, which in turn will protect and enhance the diversity of local species and increase their abundance [3].

7.3 A New Model of Integrated Development

The current development trend, when combined with the offshore wind power base, is the large-scale development of offshore wind power, the continued exploration of the deep sea, the strengthening of the deep integration of upstream and downstream industry chains, the protection and comprehensive utilization of marine resources, the conversion of environmental constraints into green opportunities, Building offshore demonstration projects that integrate different energy sources and resources, such as wind power, hydrogen energy, seawater desalination, energy storage, and marine ranches, has emerged as a new business form and model for the integrated development of offshore wind power. It also aims to develop and cultivate offshore wind power industrial clusters and support the industry's sustainable growth.

7.3.1 Offshore Wind + Ocean Ranch

In actuality, marine pastures face challenges with difficult terrain and inadequate power supplies, which make it impossible to operate and maintain large-scale contemporary resource and environmental monitoring facilities, as well as issues with high labor intensity and low efficiency. Offshore wind turbines are situated far offshore, which raises the costs of grid management and maintenance as well as causes significant power transmission losses.

The researchers combined the development of the two, taking into account the features and technical challenges faced by offshore wind farms and marine farms: spatial integration, three-dimensional development of above- and below-water areas, gathering of sea surface and sea bottom space, thorough utilization of marine biological resources and surface wind energy, to realize the spatial coupling of clean power generation and pollution-free fishery product production; structural integration to accomplish the organic integration of the artificial reef and wind turbine foundation pile through the construction of breeder wind turbine foundations, and subsequently to attain the functional integration of environmental restoration and resource conservation; In order to accomplish the integration of biological resources and wind resources in the Marine pasture yearly continuous exploitation of production time, functional integration, complete use of seasonal fishery production peak (spring, summer, autumn) and wind power peak (winter), is required.

Currently, the trial research on the integration of offshore wind power with marine aquaculture was implemented in 2000 by European countries, such as Germany, the Netherlands, Belgium, Norway, etc. [4]. The "Geng Hai No. 1" contemporary Marine ranch complex platform project, constructed by CIMC Raffles, was effectively delivered for usage in May 2020.

An important new industrial model and strategy for the growth of intensive sea use going forward is the integrated development of offshore wind generation and marine ranching. It is imperative that our nation conduct experimental research to understand the mechanisms of interaction between offshore wind power and marine pasture, as well as the effects of offshore wind power on marine pasture. We also need to develop a new integrated development model for offshore wind power and marine pasture that will enable the synchronous and efficient production of clean energy and safe aquatic products.

7.3.2 Offshore Wind Power + Wave Energy

Sea water will shift from its initial equilibrium position and flow upward, downward, forward, and backward to generate waves as sea breeze and air pressure changes. A wave is a movement that fluctuates regularly throughout time. Periodic wave energy can be converted into electrical energy using wave energy generator sets.

In order to capture wave energy, the wave energy generator set is installed on and beneath the wind turbine foundation platform, with the angle changed in accordance with the direction of the waves. Estimates place the potential utilization rate of wave energy between 60 and 80%. Furthermore, when the operation and maintenance ship is berthed, the wave energy generating set can increase safety by offsetting the energy of the platform wave in a forward direction. That's yet another significant advantage of mixing offshore wind and waves.

7.3.3 Offshore Wind Power + Hydrogen Production and Storage

The issue of how to link large-scale offshore wind farms to the grid and absorb their energy has become urgently necessary once they are operational. The new energy hydrogen production technology represented by wind power hydrogen production has progressively grown and essentially has the prerequisites for industrialization thanks to the advancement of hydrogen energy technology, particularly hydrogen production and hydrogen storage technology. As a result, surpassing the conventional understanding of hydrogen energy, hydrogen can be produced directly using offshore wind power and delivered to the market using high-pressure or liquid hydrogen storage and transportation technologies. Because the "green hydrogen" produced by offshore wind power is carbon-free, storable, transportable, and decentralized, offshore wind power may now cross power transmission channels and compete with oil and gas as a premium strategic energy source. Furthermore, because the cost of building new transmission infrastructure is considerable and using pre-made natural gas pipes can significantly lower prices, offshore wind power is becoming more and more appreciated by nations worldwide. It is also steadily expanding into the deep sea. The first offshore wind power hydrogen production project in the world is the Netherlands PosHYdon project; the Hyport Oostend offshore wind power hydrogen production project in Belgium; Rsted's 1.4GW Hornsea 2 offshore wind farm will be connected to the Gigastack hydrogen production project to produce green hydrogen; Let's hold off.

The definition of an offshore wind power-hydrogen energy integrated energy system is the utilization of sporadic, unbalanced wind power for the synthesis and storage of hydrogen. The system consists of wind power generation, a device for desalinating seawater, a device for producing hydrogen through water electrolysis, a device for storing hydrogen, an installation for fuel cell power generation, a power distribution system, associated pipelines, a monitoring system for wind turbines, and a supporting electrical access device. A hydrogen production device can be defined as one that can produce hydrogen through a combination of water electrolysis process, water electrolysis device, separator, cooler, and other equipment.

The onshore centralized control center houses the hydrogen storage and hydrogenation system, while the marine riser station integrates the hydrogen production system. The hydrogen storage system's high-purity hydrogen can be utilized as chemical building blocks to provide the system with "electric hydrogen" co-supply [5].

References

1. Tao Y, Wang Q, Xu XM. Study on underwater noise assessment and management of offshore wind farms. Adm Tech Environ Monit. 2021;33(02):1–4+8.
2. Kirchgeorg T, Weinberg I, Hörnig M, et al. Emissions from corrosion protection systems of offshore wind farms: evaluation of the potential impact on the marine environment. Mar Poll Bull. 2018;136(11):257–68.
3. Su W, Wu N, Zhang L, et al. A review of research on the effect of offshore wind power project on marine organisms. Mar Sci Bull. 2020;39(03):291–299.
4. Yang H, Ru X, Zhang L, et al. Industrial convergence of marine ranching and offshore wind power: concept and prospect. China Acad J Electron Publishing House 2019;34(06):700–707.
5. Guo M. Research on optimal operation of energy systems including hydrogen production units under integrated demand response. Shanghai Jiao Tong University; 2020.
6. Cai L, Jiang S, Ma L, et al. Review of the research on the effects of electromagnetic radiation from offshore wind power on marine organisms. Ocean Dev Manag. 2019;36(12):72–6.

Chapter 8
Offshore Wind Turbine Standards and Certification

8.1 Overview

Offshore wind turbine certification is a basic requirement for the turbine design and manufacturing industry. In the design and manufacturing process, according to relevant standards, a complete certification can supervise and control the entire process, thus guiding the healthy development of the industry [1]. Offshore wind turbine certification originated in Denmark. In 1983, Denmark implemented significant policies to encourage and support wind power, which led to the rapid expansion of its wind power industry. For a time, there was a proliferation of wind power technologies and equipment. To ensure the safe and stable operation of wind turbines, the Danish government invested heavily in supporting certification research, aiming to control the quality of wind turbine products and improve the reliability of wind turbine operation. These efforts did help regulate the wind power market further and promote the healthy and stable development of the industry [2]. Certification standards were first applied to onshore turbines even earlier with good results. However, unlike onshore turbines, offshore wind turbines face tougher external environmental conditions such as high salt fog, thunderstorms, typhoons, etc., and the hydrogeological conditions of different sea areas also differ significantly, which makes the assessment and certification more complex. Therefore, establishing regulations suitable for offshore turbines is crucial for the development of offshore wind power.

M. Ge et al., *Offshore Wind Turbine Technology*,
https://doi.org/10.1007/978-981-96-2175-0_8

8.2 Offshore Wind Turbine Standards

8.2.1 *Common Standard*

Offshore wind turbine standards were initially developed and implemented in Denmark, Germany, and the Netherlands. As the importance of certification became widely recognized among manufacturers, suppliers, investors, operators, as well as financial and insurance institutions, other countries, including the United Kingdom, the United States, Greece, India, Spain, and China, has also established their own certification systems to address specific national requirements [2, 3]. Some of the most commonly referenced offshore wind turbine certification standards include IEC61400-3, GL Offshore Wind Guidelines, Danish Recommendations, DNV-OS-J101, and IEC WT01. The specific scope and application of these standards are outlined in Table 8.1 [1–3].

1. **IEC61400-3**

IEC 61400-3, first issued in 2005, sets forth the minimum requirements for the design of offshore wind turbines. It provides criteria for evaluating external conditions at offshore wind farm sites, ensuring the structural integrity of turbine designs. The standard's primary objective is to offer adequate protection against various hazards throughout the turbine's operational life. However, it is not a comprehensive design or installation guide and must be used in conjunction with other relevant IEC/ISO standards. Notably, it fully aligns with the requirements of IEC 61400-1, and includes relevant content in clauses that need detailed descriptions [4].

As part of the broader IEC 61400 series, IEC 61400-3 focuses on design, safety, and testing requirements for wind turbines. While it provides detailed guidance on site evaluation and load assumptions, other critical areas, such as materials, structural components, mechanical parts, and systems like safety and electrical systems, are only briefly addressed or omitted. In terms of material standards, IEC 61400-3

Table 8.1 Application scope of offshore wind turbine standards

Standards	Project certification	Loads	Support structures	Mechanical components	Safety, electrical and CMS
IEC61400-3		Y	(Y)		(Y)
GL Offshore Wind Guidelines	Y	Y	Y	Y	Y
Danish Recommendations		Y	Y		
DNV-OS-J101		Y	Y		
IEC WT01	Y	(Y)			

Note Y denotes that the standard can address the subject independently; (Y) denotes that the standard can only partially address or is dependent on other standard offers guidance similar to that in IEC 61400-1, recommending the adoption of relevant national and international standards for determining key elements of turbine structural integrity. Care must be taken when using safety factors from these external standards alongside those in IEC 61400-3 to ensure the overall safety level is not compromised [3].

In summary, IEC 61400-3, in conjunction with national and international design codes, defines load assumptions and safety levels to ensure the integrity of turbine structures, mechanical components, blades, and electrical systems [3].

2. **GL Offshore Wind Guidelines**

In 1995, Germanischer Lloyd (GL) published the first certification guidelines for offshore wind turbines. Later, in 2005, based on extensive experience in the design, certification, and operation of offshore wind farms, a more complete version was revised. Further, in 2012, the guidelines were refined to form a final version. The GL Offshore Wind Guidelines provide comprehensive design requirements for the entire wind turbine, covering all aspects of certification procedures, including loads, materials, structures, mechanical components, rotor blades, electrical safety, and operational condition monitoring [3].

The safety system in this guideline is based on the expertise and knowledge of onshore wind turbines. That is, the load safety factors are consistent with the IEC 61400-1 and IEC 61400-3 standards, and the material safety parameters are also similar to those in IEC standards but more specific. The IEC standards specify material safety factors in a general way and do not consider the specific material itself. In contrast, the GL standards take into consideration the mechanical properties of the materials when determining safety factors. For instance, compared to welded structures, the safety factors for soil resistance are more conservative, as there is considerable uncertainty in determining soil resistance [1, 3].

The scope of type certification and project certification is also outlined in the GL Offshore Wind Guidelines. Project certification is divided into Class A and Class B certifications. Users can choose to monitor all wind turbines in the wind farm (100%) for Class A certification or randomly select one out of every four turbines, i.e., monitoring 25% of the wind turbines for Class B certification. This means that for project certification, at least 25% (Class B) of the wind turbines in a wind farm must be supervised and monitored by a third party [2].

3. **Danish Recommendations**

In 2001, Denmark introduced mandatory technical certification regulations for offshore wind turbines as an annex to the "Danish Technical Standards for Wind Turbine Type Approval and Certification" and the "Regulations for Structural Safety and Load of Wind Turbine" (DS 472). These regulations provide detailed descriptions and supplementary information on the technical requirements necessary for offshore wind turbine approval, with a primary focus on load-bearing and foundation-related concerns. Additionally, they address areas such as lightning protection, labeling, noise emissions, and environmental impact assessments. However, the Danish guidelines do not establish specific offshore requirements for mechanical, electrical, or safety systems [1, 2, 5].

4. **DNV-OS-J101**

In 2004, the Det Norske Veritas (DNV) issued the first standard for the structural design of offshore wind turbines. This standard only covers the support structures of wind turbines and is applicable to all structures below the nacelle, including foundations and soil. The design principles and requirements address aspects such as reinforced concrete structures, grouted connections, corrosion protection, load effects, transportation, and installation, etc. However, it does not completely cover the entire system that should be considered in the type certification of offshore wind turbines [3], such as blades, mechanical components, electrical equipment, safety systems, and operational condition monitoring, etc.

5. **IEC WT01**

IEC WT01 outlines the basic principles and procedures for wind turbine certification. Although it does not provide any design requirements and does not include offshore wind turbines, its definition of project certification has been adopted in the project certification of some offshore wind farms. This is mainly due to the fact that there is currently no other standard that establishes a certification system for offshore wind turbines, and that IEC WT01 provides a "convenient project certification" process. In summary, IEC WT01 is an international standard that defines the standards and procedures for type certification and project certification. The GL Offshore Wind Guidelines also incorporated the advantages of WT01 in project certification and additional supervision for offshore wind turbines. At present, IEC WT01 has been revised and improved to include offshore wind farm certification. It is suitable for third-party supervision during the manufacturing, transportation, installation, and commissioning of wind turbines, as well as for the evaluation of overall project and design documentation [5].

8.2.2 *Challenges of Offshore Wind Turbine Standards*

1. **Complex Loading Conditions of Offshore Wind Turbines**

The external environmental conditions of offshore wind turbines are crucial for determining design loads. Similar to onshore wind turbines, offshore wind turbines include normal load cases, extreme load cases, special load cases, and transport load cases. A key difference is that offshore wind turbines also experience unique wave load cases [1]. The loads acting on offshore wind turbines typically include wind, wave, and ice loads, which primarily affect the support structure. Additionally, the turbines are subject to minor loads from ocean currents. Preliminary analysis indicates that wind and wave loads are the two main sources of load in most cases. This nonlinear, multi-physics coupling process presents significant challenges in load calculations for offshore wind turbines. In certain regions, such as the northern Baltic Sea, ice combined with strong winds is the dominant external factor considered in turbine design. However, it is generally believed that ice and waves will not occur simultaneously in the near-shore areas. When designing offshore wind turbines, it is important to follow the principle that "the less information available about the real environment, the more conservative component in the load assumptions should be considered" [5].

2. **Overall Design of Large-Scale Offshore Wind Turbines**

The overall design of large-scale offshore wind turbines faces significant challenges. Compared to onshore turbines, the installation and maintenance of offshore wind turbines are more difficult, and require a separately designed foundation structure. To reduce engineering costs and lower the levelized cost of energy (LCOE) over the lifecycle, the capacity of offshore wind turbines is much higher than that of onshore turbines. Earlier offshore wind farms commonly adapted modified onshore turbines for offshore use, but serious failures often occurred. As offshore wind farms become larger and are located in deeper waters, turbines with higher capacity have seen rapid development and wide application. However, when the rotor diameter exceeds 200 m, there is a critical challenge for the safety of the rotor and other components, including issues with aerodynamics, structural design, transportation, installation, operation, and maintenance. Therefore, new engineering solutions are urgently needed to consider all aspects of wind turbine design and analysis, and existing standards and guidelines should be further improved [3].

3. **Complex External Conditions of Offshore Wind Turbines**

The offshore wind turbines face complex external environmental conditions, with challenges such as typhoons, thunderstorms, and salt fog, and the geological conditions of the seabed are ever-changing. These factors place higher demands on the safety standards, typhoon resistance, lightning protection, anti-corrosion technologies, ventilation, and cooling systems, as well as scouring protection techniques of offshore wind turbines, which has also presented difficulties for the improvement of relevant standards.

8.3 Offshore Wind Turbine Certification

8.3.1 Development of Offshore Wind Turbine Certification

Wind turbine certification is an essential measure to ensure the quality and reliability of offshore wind power equipment. With a history of nearly 35 years, it has played a significant role in advancing the global wind power industry. The international wind power sector, particularly in Europe, has developed a comprehensive set of technical standards for both complete turbines and their components. This includes a certification system that covers design evaluation, quality management assessments, manufacturing supervision, and prototype testing, providing technical safeguards for the manufacturing and procurement of wind power equipment. Initially, only Denmark, Germany, and the Netherlands implemented mandatory offshore wind certification, and these countries remain leaders in this field. In recent years, countries like Greece, India, Spain, Sweden, the United States, and China have recognized the need for comprehensive evaluations and certifications of wind turbines and their installation processes. Third-party certifications are commonly adopted in these countries to assess the design, implementation, and production of such projects. Numerous third-party organizations are playing important roles in wind turbine certification internationally, including GL WIND, TÜV-NORD, TÜV-SÜD, BSI, EV, and SGS [1, 5].

In the process of offshore wind turbine certification, the relevant definitions are as follows:

(1) Scope of Certification

This includes type certifications for offshore wind turbine (entire systems), critical components (such as blades, gearboxes, generators, foundations, towers, etc.) and project certification for offshore wind farms.

(2) Certification Basis

The certification is based on relevant national and industry standards, as well as applicable international standards.

(3) Certification Body

The Wind Power Certification Council authorizes and commissions the Certification Center to conduct type certification and project certification for offshore wind turbine. The Offshore Wind Equipment Testing Center and the Technical Committee provide technical support for offshore wind turbine certification.

(4) Certification Methodology

The certification process for offshore wind turbine includes both type certification and project certification. Type certification involves assessment of new wind turbine models against standards and specifications through product design evaluation, production installation assessment, quality management system review, and prototype testing. Project certification evaluates whether offshore wind turbine and critical components, already type-certified, are suitable for environmental factors and meet other requirements related to the wind farm site. The certification body must assess whether the wind resources, ocean conditions, grid requirements, and soil characteristics at the site are consistent with the parameters specified in the design documents of the certified offshore wind turbines and components. During project certification, the offshore wind turbine/ wind farm will be monitored throughout manufacturing, transportation, installation, and trial operation, with regular inspections conducted at fixed intervals.

Certification is a necessary and crucial step to ensure the safety and reliability of wind turbines. The transition from onshore to offshore brings new challenges to the certification process, requiring not only robust certification technical capabilities but also extensive offshore technical expertise [1, 5].

8.3.2 *Type Certification for Offshore Wind Turbines*

1. **Steps in Type Certification**

In type certification, the overall design of the offshore wind turbine should be assessed. Type certification covers all components of the wind turbine, including the inspection, evaluation, and certification of its safety, design, structure, manufacturing processes, and quality. The key steps and aspects in type certification include product design evaluation, production and installation requirements, quality management system audits, and prototype testing. Once the evaluation is passed, a type certification certificate is issued, as shown in Fig. 8.1 [5].

Design evaluation assesses the safety and reliability of a product. It involves the approval of design drawings and technical documents, and/or the operational testing of wind turbines and critical components to verify that the design meets relevant codes, standards, and applicable technical conditions. If the design drawings and technical documents are approved, or if the operational testing of wind turbines and critical components is successfully validated and meets the relevant requirements, a design evaluation certificate will be issued.

During the production and installation phases, the implementation of design requirements refers to the evaluation of the factory's production conditions, inspection procedures, and manufacturing processes. This, combined with the results of type testing or inspections, assesses the factory's ability to ensure product quality within the scope of its manufacturing application. The evaluation ensures that the factory can produce products that meet relevant standards and that the requirements

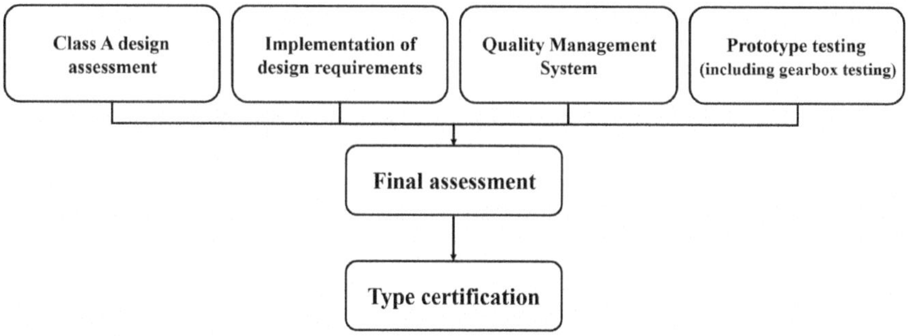

Fig. 8.1 Steps in type certification

outlined in the technical documents for components are adhered to during production and installation.

Within the scope of the Quality Management (QM) system, manufacturers must demonstrate that their design and manufacturing processes meet the ISO 9001 requirements, which is generally achieved through certification by an accredited certification body.

Prototype testing typically includes power curve measurement, noise measurement, electrical performance testing, wind turbine characteristic testing, and load measurement.

2. **Main Contents of Certification Testing**

(1) Power Characteristics Testing

- To obtain the power characteristic curve, it is necessary to estimate the annual average power output of the offshore wind turbine.
- The test content includes environmental parameters such as wind speed, wind direction, atmospheric pressure, atmospheric temperature, air density, relative humidity, as well as operational parameters of the offshore wind turbine such as power output.
- Testing should be conducted according to GB/T 18451.2 or relevant standards. After testing, the measured power characteristic curve should be compared with the design power curve, with particular attention to whether it aligns with the assumed rated wind speed and rated power. If there are changes to the rotor, blade type, or rotor diameter, the power characteristics must be retested; if there is a deviation in the rated rotational speed of the rotor, retesting is also required.
- The power characteristic curve includes the following: standard power curve; power coefficient curve; scatter plot of power under wind turbulence; table of power characteristic following wind speed distribution; theoretical annual energy production of the wind turbine calculated by hub-height average wind speed based on the standard power curve; daily temperature variations during testing; and test error analysis.

(2) Load Testing

- Load testing involves loads measurement under design calculations and specific conditions. The test should be conducted on wind turbines with similar dynamic characteristics and structure as the ones submitted for certification, allowing possible differences in detail. To reduce errors, the applicant should provide the predicted load and dynamic characteristics of the wind turbine being tested.
- The test content includes blade root loads (flapwise and edgewise bending moments); rotor loads (pitch moment, yaw moment, and torque); tower loads (base bending moments in two directions), and if necessary, top bending moments; meteorological parameters including wind speed, wind direction, air temperature, and air density, with additional measurements of wind shear and temperature gradient

if needed; operational parameters of the offshore wind turbine, such as electrical power, rotor speed, pitch angle, yaw position, rotor azimuth, and grid conditions.
- Testing should be conducted according to IEC 61400-13 or relevant standards. In addition to the requirements in IEC 61400-13, all loads related to sea conditions should also be tested to verify the effects on the wind turbine.
- Test results include data description tables; measurement data logs; time series curves showing wind speed, blade loads, rotor loads, and tower loads; changes in blade flapwise and edgewise bending moments and main shaft torque with rotor azimuth; load spectrum analysis charts; fatigue load spectra; wind speed measurement statistics; operational parameter statistics; load statistics; fatigue load statistics and etc.

(3) Noise Measurement

- Noise characteristics of the offshore wind turbine during operation, including environmental effects, should be confirmed. Protective measures should be implemented based on the noise level, and the effectiveness of noise reduction and protection should be verified. Environmental noise should comply with national laws and regulations regarding its impact on nearby residents. The noise emission characteristics of offshore wind turbines are determined through appropriate testing and analysis.
- Test content includes sound intensity and sound pressure level at a wind speed of 8 m/s; 1/3 octave band sound pressure; tonal noise; sound propagation directions from three fixed locations; noise frequencies above the minimum threshold, etc.
- Testing should be conducted according to IEC 61400-11 or relevant standards, with particular attention to the rotor and blade types, tower height, tower type, and gearbox type (if applicable).
- Test results include calculation of sound pressure levels at corresponding test points in the wind farm, along with noise assessment; potential tonal noise in the narrowband spectrum at non-nominal sound pressure levels.

(4) Power Quality Testing

- The capacity of the offshore wind turbine used for power quality testing should be significantly smaller than the grid connected to.
- Test content includes:

  Long-term tests, including the relationship between active power and wind speed, reactive power and wind speed, the relationship curve between active and reactive power, and voltage variation during power generation;

  Short-term tests, including the voltage variation, current fluctuation, harmonics, voltage flicker, and inrush current. Depending on the situation, additional tests may be required, such as environmental tests for electric field distribution and electromagnetic compatibility (EMC) tests.

- Testing should be conducted according to GB/T 20320 or relevant standards. If there are changes in generator type or inverter type (if applicable), retesting is required.
- Test results include grid power supply conditions; on-site wind resource variations; active power and wind speed curves; reactive power and wind speed curves; corresponding curves of active and reactive power; voltage (including flicker) and current fluctuation curves; harmonic content variation curves and etc.

3. **Main Contents of Certification Performance Assessment**

(1) Power Curve

During the operational phase of a wind farm, the IEC 61400-12 standard is typically used to evaluate the power curve of wind turbines, which includes the following sections:

- IEC 61400-12 Part 1: Power performance measurement of wind turbines.
- IEC 61400-12 Part 2: Power performance verification of each wind turbines based on the nacelle-mounted anemometers.

Most turbine supply agreements (TSAs) specify that within one week after the turbine acceptance tests, a certain number of turbines in the wind farm must undergo power curve verification according to IEC 61400-12 Part 1. The number of turbines ranges from one to 5% of the total, and the verification period can range from 3 months to the entire warranty period, up to a maximum of 5 years. On-site calibration is also required based on the site conditions.

The power characteristics of wind turbine are determined by the standard power curve and the estimated annual energy production (AEP). The standard power curve is derived from synchronized wind speed and power output data collected during on-site testing, and adjusted to standard conditions. The data collection period must be long enough to establish a statistically significant database covering the full range of wind speeds, wind conditions, and atmospheric conditions. AEP is calculated by applying the standard power curve to the given wind speed frequency distribution, assuming 100% availability. A deviation between the AEP with the reference energy in TSA is generally up to 5%.

Measurement of the wind turbine power curve according to IEC 61400-12 Part 1 is time-consuming and expensive, and is typically applied to a limited number of turbines in a specific wind farm, of which the turbines should be representative. However, IEC 61400-12 Part 2 can be used to verify the power curve of a single wind turbine. This method adopts nacelle-mounted anemometer data as the wind speed source instead of data from the meteorological masts, making the process simpler.

(2) Availability

① Definition and Calculation of Availability

Availability refers to the percentage of time that a wind turbine is operational and fault-free during the assessment period. Time lost due to shutdowns caused by conditions beyond the control of the wind turbine design, such as grid failures or curtailment requests, is not included. Typically, the availability is evaluated and calculated on an annual basis.

The formula of availability of a single wind turbine is

$$A(wt) = \frac{T_t - [D - (T_g + T_w + T_o)]}{T_t} \times 100\% \tag{8.1}$$

where $A(wt)$ is the availability of a single wind turbine; T_t is the total duration of the assessment period; T_g is the duration of grid outages during the assessment period; T_o is the other downtime; T_w is the duration when wind conditions or other environmental factors exceed the specifications; D is the total downtime during the assessment period.

T_o covers all exceptional circumstances, such as scheduled maintenance, force majeure, owner's instructions and risks, lightning damage, icing, restricted access, communication failures, and cable uncoiling.

Additionally, if any operational or status parameter of the wind turbine cannot be displayed or detected by the monitoring system, there may be a hidden fault, causing the turbine to continue running in a compromised state. If this is confirmed, the corresponding operational time should be considered downtime due to fault and included in parameter D.

The average availability of the wind turbine is

$$A(wf) = \frac{1}{N} \sum_{i=1}^{N} A(wf_i) \tag{8.2}$$

where $A(wf)$ is the average availability of wind turbines; $A(wf_i)$ is the availability of a single wind turbine during the assessment period; N is the number of turbines in the wind farm.

② Evaluation of Wind Turbine Availability

In China, wind turbine manufacturers are typically required to guarantee an availability of no less than 95% during the warranty period, with the annual availability of individual turbines not falling below 85%. The guaranteed availability values should be clearly stipulated in the contract; if the guaranteed value is not met, the manufacturer shall pay liquidated damages as compensation for performance degradation. The formula of the liquidated damages is

$$LD = \left[E \cdot A(wwf) - E \cdot A(wf)\right] \cdot EP \tag{8.3}$$

where LD is the liquidated damages; E is the theoretical power generation of the wind farm during the assessment period, calculated based on the P50 power generation level, and taking all loss factors into account; $A(wwf)$ is the guaranteed average availability of wind turbines; $A(wf)$ is the average availability of wind turbines; EP is the average electricity price during the assessment period.

Similarly, if the wind turbine manufacturer improves management, technology, and service levels, resulting in the actual availability of the wind turbine exceeding the guaranteed availability, the wind farm owner should share the increased energy revenue with the manufacturer. This has become common practice in foreign markets. Typically, the share of increased energy revenue between the wind farm owner and the wind turbine manufacturer ranges from 30 to 40% of the improvement in availability.

③ Evaluation of Availability Based on Power Output

When conditions permit, it should be considered to assess availability based on power output, specifically evaluating the Loss of Production Factor (LPF).

In this context, power loss refers to the amount of power that the wind turbine fails to produce during downtime, calculated based on the power output of other nearby wind turbines. The formula of the loss of production factor for a wind turbine is

$$LPF_t = \frac{EL_t}{EP_t + EL_t + EC_t} \times 100\% \tag{8.4}$$

where LPF_t represents the power loss of the assessed wind turbine related to the manufacturer's responsibilities during the assessment period; EL_t is the power loss due to downtime of the assessed wind turbine directly related to the manufacturer's responsibilities, including maintenance services, turbine failures, controller interruptions, and shutdowns; EP_t is the cumulative power output of the wind turbine during the assessment period; EC_t is the downtime power loss of the assessed wind turbine unrelated to the manufacturer's responsibilities, which includes losses due to grid conditions, climatic factors, curtailment by dispatch, and other reasons attributed to the owner.

The formula of the power loss rate of the wind farm is

$$LPF_f = \frac{1}{N} \sum_{i=1}^{N} LPF_{ti} \tag{8.5}$$

where LPF_f represents the power loss rate of the wind farm related to the manufacturer's responsibilities during the assessment period; N is the number of wind turbines in the assessed wind farm; $i = 1, 2, \cdots, n$; LPF_{ti} is the power loss rate of the i-th wind turbine related to the manufacturer's responsibilities. Typically, wind turbine manufacturers can be required to guarantee that the power loss rate of the wind farm, LPF_f, does not exceed 5% during the warranty period.

(3) Reliability

Unlike availability, the reliability of a wind turbine is directly related to the quality of its design, production, installation, commissioning, and maintenance. A reasonable setting of reliability parameters and indicators is key to accurately reflecting the quality of the wind turbine.

Currently, China National Energy Administration (CNEA) has issued and implemented the "Reliability Evaluation Regulations for Power Generation Equipment" and the "Reliability Evaluation Regulations for Wind Power Equipment (Trial)". International reliability evaluation methods for wind power are also worth considerations.

In recent years, most major wind turbine manufacturers abroad have been tracking the Mean Time Between Inspections (MTBI) of their turbines and are committed to improving this metric, with reported values ranging from 40 to 70 days, compared to only 28 days in reports from 2000 to 2002.

MTBI refers to the number of days between two scheduled or unscheduled maintenance events. To accurately reflect the reliability performance of wind turbines, MTBI needs to be calculated as follows:

- Calculations must be conducted based on the type and model of the wind turbine and span a sufficiently long period (at least 6 months);
- Switching the maintenance key of the wind turbine controller to "LOCAL" counts as one maintenance event;
- If the maintenance key is switched multiple times in one day, it is counted as only one maintenance event;
- If a single maintenance event lasts several days, it is still considered one event;
- Mandatory and owner-requested inspections are not included.

Measuring and calculating MTBI can not only evaluate the reliability of wind turbines but also assess the efficiency of their maintenance. Some specific parameters are as follows:

- Mean Time Between Failures (MTBF): A direct measure of the design quality of the wind turbine and the reliability of its components;
- Mean Time to Repair (MTTR): A direct measure of the efficiency of the maintenance service team's fault diagnosis and repair, including adequate spare parts availability;
- Planning Capability of the Maintenance Team: The ability of the maintenance team to plan and bundle sub-tasks within the same maintenance project, which can effectively reduce the number of maintenance events and improve MTBF.

MTBI is a important parameter for assessing the performance of wind turbines. Currently, the MTBI evaluation method has not been adopted in the Chinese wind power industry. With the rapid growth of installed wind power capacity and the continuous expansion of wind turbine production capacity, reliability issues are becoming increasingly prominent. Therefore, it is recommended to include MTBI as a key parameter for assessing wind turbine reliability. By strengthening the management and supervision of operational information from wind farms and establishing post-evaluation mechanisms and quality supervision systems in the wind power industry, it is believed that in the near future, MTBI will become an important basis for evaluating the quality of wind power products, just as it has in foreign wind power industries.

(4) Grid Compatibility

With the rapid growth in installed wind power capacity, the ability of the grid to integrate wind power has become a significant factor affecting its development. Therefore, wind turbines are required to have good grid compatibility, being able to adjust active and reactive power according to grid requirements. In the event of brief grid faults, the turbines should remain connected and support grid operation. Specific requirements should comply with the officially released national standard "Technical Regulations for Wind Farm Connection to Power Systems" (revised edition).

(5) Noise Curve

The noise curve of a wind turbine describes the relationship between noise level (measured in dB(A)) and wind speed. The rated noise level of the wind turbine is measured at a height of 10 m with a wind speed of 8 m/s.

Pitch-controlled or active stall wind turbines can restrict pitch activities within a certain envelope, achieving different noise levels at 8 m/s, which is typically used to reduce the impact of wind farm noise on nearby residential areas. However, this may affect the power curve and power generation. Therefore, modern wind turbines need to provide an optimal balance between power curve and noise impact. Manufacturers are generally required to provide power curves at different noise levels, preferably certified by a recognized third-party organization, for evaluation during the procurement phase. The noise curve of wind turbines is usually measured according to the "Noise Measurement Techniques" in IEC 61400-11 standard [5].

8.3.3 Project Certification for Offshore Wind Turbines

The project certification of offshore wind turbine involves strict monitoring and management of the entire lifecycle of offshore wind power projects, from design and construction to operation. This process covers all aspects, from siting and planning of wind farms to the design, manufacturing, installation, commissioning, and even the operation of equipment, providing comprehensive certification services across the entire wind power industry chain. Offshore wind turbine project certification involves numerous technical elements, and the ability to perform such certification fully demonstrates the technical capabilities of certification bodies. Generally, the offshore wind turbine project certification follows the steps outlined in Fig. 8.2.

1. **Site Assessment**

The site assessment includes examining the impact of environmental factors on offshore wind turbines and analyzing the interactions between offshore wind farms, in which the following factors need to be considered:

- Wind conditions;

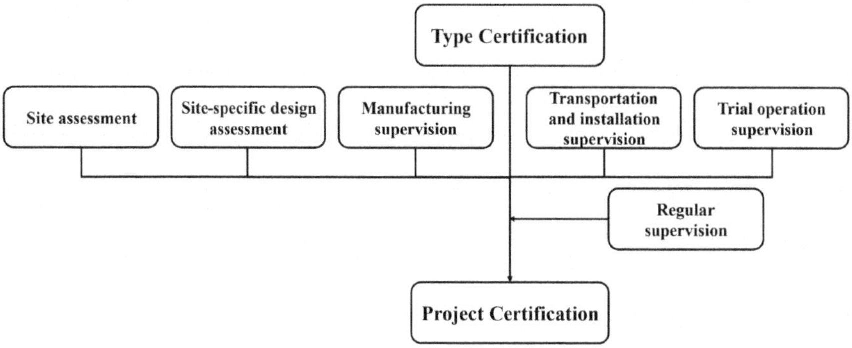

Fig. 8.2 Steps in project certification

- Ocean conditions, including sea depth, waves, tides, sea ice, seabed erosion, and etc.;
- Engineering geological conditions, including topography, geomorphology, composition and characteristics of the foundation soil, earthquake effects at the site, and etc.;
- Site location and wind farm layout;
- Other environmental conditions, such as salt content in the air, temperature, snow and ice, humidity, lightning strikes, solar radiation, and etc.;
- Grid conditions.

These site conditions will be assessed based on the following aspects: the rationality and completeness of the measurement report; conformity to the standard requirements; the qualifications of the external environmental measurement agency; and etc.

2. **Site-Specific Design Assessment**

The site-specific design assessment can be detailed as the following steps:

- Assumptions of site-specific loads;
- Comparison of site-specific loads with loads in the type certification;
- Site-specific support structures (tower, underwater structure, and foundation);
- Residual stress calculation of mechanical parts and rotor blades (if loads exceed those certified in the type certification);
- Modifications to mechanical parts and rotor blades related to type certification (if any modifications are made).

3. **Manufacturing Supervision**

Before manufacturing supervision, the manufacturer should meet certain quality management requirements, typically the certification of a quality management system according to ISO 9001.

The scope of manufacturing supervision and the number of measurement samples depend on the quality management measurement standards. Generally, the following operations and approvals are required:

- Inspection and testing of materials and components, including document and record review, on-site verification, and interviews with management;
- Detailed review of quality management records, such as quality management reports, type certification certificates, certification body, and certificate validity;
- Supervision of manufacturing processes, including storage conditions, methods, and random sampling;
- Corrosion protection inspection;
- Measurement of dimensions and tolerances;
- General inspection of appearance and damages.

4. **Transportation and Installation Supervision**

Before the transportation and installation supervision, transportation and installation manuals should be submitted, which may consider the site-specific environmental conditions if necessary. These manuals must fully account for the specific transportation conditions, equipment characteristics, and the key installation conditions of the site, such as climate and work schedule.

The scope of supervision and the number of samples to be measured depend on the quality management standards of the transportation and installation company. The typical tasks include:

- Approval of transportation and installation procedures;
- Identifying and locating all questionable turbine components;
- Checking for damaged components during transportation;
- Review of the work schedule, including welding, installation, concrete pouring, and bolt tightening;
- Verifying the pre-processed parts and components to be installed meet production quality requirements;
- Random supervision of key steps in the installation process, such as piling, concrete pouring, and wind turbine installation;
- Inspection of concrete pouring and bolt connections, and supervision of non-destructive testing;
- Inspection of corrosion protection measures;
- Inspection of anti-erosion protection systems;
- Inspection of electrical installations, including wiring, equipment grounding, and grounding systems;
- Inspection of seabed fastenings and offshore operations.

5. **Trial Operation Supervision**

The trial operation supervision covers all turbines in the offshore wind farm, ensuring that they can operate in compliance with all standards and requirements.

Before the trial operation, startup manuals and test plans must be submitted. The manufacturer should also provide evidence that the offshore wind turbines have been properly installed and tested as thoroughly as possible to ensure safe operation. If such evidence is lacking, necessary tests should be conducted before the wind turbines are put into operation.

The supervision process includes on-site observation of about 10% of the offshore wind turbines by inspectors during actual trial operation, while the remaining turbines will be inspected after the trial run. During the trial, the operation and all safety function modes of turbines will be tested. This process involves the following tests and operations:

- Function of the emergency stop button;
- Activation of the brake system under all possible operating conditions;
- Function of the yaw system;
- Turbine performance in loss of load, overspeed, and automatic operation modes;
- Visual inspection of the entire offshore wind turbine;
- Logical check of control system indicators.

6. **Regular Supervision**

To maintain the validity of the certificate, maintenance of the offshore wind farm should be carried out in accordance with the approved maintenance manuals, and the condition of the wind turbines should be regularly monitored. The maintenance must be performed by authorized personnel and documented. The intervals for regular monitoring will be specified in the inspection plan, and may vary depending on the condition of the wind turbines.

The content of regular supervision includes foundation and scour protection (if applicable, only a detailed review of relevant inspection records is needed); foundation; tower; nacelle; all parts of the transmission system; rotor blades; hydraulic/pneumatic system; safety control system; electrical installations; and etc.

7. **Project Certification: Class A and Class B**

Project certification will be issued after successfully completing all the steps described above. According to the GL Offshore Wind Guidelines, project certification is divided into two classes:

Project Certification Class A: Supervising 100% of the offshore wind turbines, meaning all turbines in the wind farm are monitored. The supervision covers key parts of the support structure, mechanical components, blades, and electrical systems.

Project Certification Class B: Supervising 25% of the wind turbines by random sampling, meaning that at least one-quarter of the turbines are monitored. The supervision includes key parts of the support structure, mechanical components, blades, and electrical systems. If significant issues, deviations from the certified design, or quality management problems are found, the number of turbines under supervision will be doubled.

8. **Re-certification**

After the expiration of the type certification, the manufacturer shall apply for re-certification. Upon completion of this process, GL WIND will issue a new type certification along with a re-certification statement, valid for two years.

For re-certification, the following documents should be submitted to GL WIND for evaluation:

- A list of valid drawings;
- A list of component design modifications from the design assessment and the assessment documentation for the modifications (if applicable);
- A list of changes in the quality management system since the latest audit;
- A list of all installed wind turbines of this model (at least including detailed model names, serial numbers, hub heights, and installation locations);
- A list of all damage records for wind turbines installed.

If structural changes have been made, GL WIND will review them and issue a revised compliance statement for Class A design assessment [5].

8.4 Risk Assessment

The risk assessment includes all risks related to the planning, development, approval, and operation of wind farms. In the assessment process, due diligence is considered as a key factor, primarily assessing technical, economic, and legal risks based on all detailed standards. Via project control methods, the feasibility of the project should be demonstrated to investors, financiers, or other stakeholders, covering technical control, financial viability, project partnerships/contract qualifications, and the interfaces between them [3].

To ensure the safety of the shipping industry, it is necessary to study the impact of offshore wind farms and assess the risks related to traffic and the environment.

During the approval process, a detailed risk analysis should be submitted to allow the responsible regulatory agencies to evaluate and minimize risks based on a comprehensive understanding of the standards. The risk analysis provides essential safety information for insurance companies and wind farm operators.

During the approval period, the regulatory agencies should set permissible values for collision frequency and risks. With the calculations of the collision frequency and risk values, an evaluation can be conducted, and the approval decision can be made only after the comparison between the permissible values and the calculated values. This not only verifies the effectiveness of risk mitigation measures but also directly influences the acceptance of the wind farm planned [3].

References

1. Wu JL, Li CF. Design of offshore wind turbines. Beijing: Chemical Industry Press; 2011.
2. Fang T, Song QY. Research on certification of wind power generation. Yizhong Technol. 2013;3:71–4.
3. Taylor Widdell G. Offshore wind energy. Beijing: Ocean Press; 2012.
4. Wind Turbines—Part 3: design requirements for offshore wind turbines: IEC 61400-3.
5. Yuan Y, Yan HM, Zhang G, et al. Offshore wind power generation technology. Nanjing: Hohai University Press; 2014.

The manufacturer's authorised representative in the EU is Springer Nature Customer Service Centre GmbH, Europaplatz 3, 69115 Heidelberg, Germany. If you have any concerns regarding our products, please contact ProductSafety@springernature.com

Printed and bound by CPI Group (UK) Ltd, Croydon, CR0 4YY
01/10/2026
02225926-0004